Internet Dating

Internet Dating deals primarily with the experiences of UK and Australian daters, examining their online accounts to see what kinds of narratives, norms, emotions and 'chemistry' shape their dating.

Has the emergence and growth of internet dating changed the dating landscape for the better? Most commentators, popular and academic, ask whether online dating is more efficient for individuals than offline dating. We prefer a socio-political perspective. In particular, the book illustrates the extent to which internet dating can advance gender and sexual equality. Drawing on the voices of internet daters themselves, we show that internet dating reveals how social change often arises in the unassuming, everyday and familiar.

We also pay attention to often ignored older daters and include consideration of daters in Africa, Scandinavia, South America, Asia and the Middle East. Throughout, we explore the pitfalls and pleasures of men and women daters navigating unconventional directions towards more equitable social relations.

Chris Beasley is Emerita Professor and Fellow of Academy of the Social Sciences in Australia, Department of Politics and International Relations, University of Adelaide. Her recent book is *The Cultural Politics of Contemporary Hollywood Film: Power, Culture and Society* (2019), co-authored with Heather Brook. Her research interests include socio-political theory, gender, sexuality, care, intimacy, men and masculinities studies, cultural politics and popular film.

Mary Holmes is Professor of Emotions and Society, Department of Sociology, University of Edinburgh. Her most recent book is *Young Refugees and Forced Displacement: Navigating Everyday Life in Beirut* (2020), co-authored with Riga, L., Holmes, M., Dakessian, A., Langer, J. and Anderson, D. Her research interests include emotions, intimacy, heterosexuality and gender.

Routledge Advances in Sociology

307 The Politics of Europeanisation
Work and Family Life Reconciliation Policy
Nazli Kazanoglu

308 Internet Dating
Intimacy and Social Change
Chris Beasley and Mary Holmes

309 Transcending Modernity with Relational Thinking
Pierpaolo Donati

310 Exploring Welfare Bricolage in Europe's Superdiverse Neighbourhoods
Jenny Phillimore, Hannah Bradby, Tilman Brand, Beatriz Padilla and Simon Pemberton

311 The Home in the Digital Age
Antonio Argandoña, Joy Malala and Richard C. Peatfield

312 Coronavirus Capitalism Goes to the Cinema
Eugene Nulman

313 Suicide Social Dramas
Moral Breakdowns in the Israeli Public Sphere
Haim Hazan and Raquel Romberg

314 Understanding China through Big Data
Applications of Theory-oriented Quantitative Approaches
Yunsong Chen, Guangye He and Fei Yan

For more information about this series, please visit:
https://www.routledge.com/Routledge-Advances-in-Sociology/book-series/SE0511

Internet Dating

Intimacy and Social Change

Chris Beasley and
Mary Holmes

Routledge
Taylor & Francis Group
LONDON AND NEW YORK

First published 2021
by Routledge
2 Park Square, Milton Park, Abingdon, Oxon OX14 4RN

and by Routledge
605 Third Avenue, New York, NY 10158

Routledge is an imprint of the Taylor & Francis Group, an informa business

British Library Cataloguing-in-Publication Data
A catalogue record for this book is available from the British Library

Library of Congress Cataloging-in-Publication Data
Names: Beasley, Chris, author. | Holmes, Mary, 1965– author.
Title: Internet dating : intimacy and social change / Chris Beasley and Mary Holmes.
Description: 1 Edition. | New York : Routledge, 2021. | Includes bibliographical references and index.
Identifiers: LCCN 2020049575 (print) | LCCN 2020049576 (ebook) | ISBN 9780415720694 (hardback) | ISBN 9781315866796 (ebook)
Subjects: LCSH: Dating (Social customs)—Computer network resources. | Personals—Computer network resources. | Internet.
Classification: LCC HQ801.82 .B43 2021 (print) | LCC HQ801.82 (ebook) | DDC 306.730285—dc23
LC record available at https://lccn.loc.gov/2020049575
LC ebook record available at https://lccn.loc.gov/2020049576

ISBN: 978-0-415-72069-4 (hbk)
ISBN: 978-0-367-75393-1 (pbk)
ISBN: 978-1-315-86679-6 (ebk)

Typeset in Times New Roman
by Apex CoVantage, LLC

Access the Support Material: www.routledge.com/9780415720694

We would like to dedicate this co-authored book to the third musketeer who for so many years was our other co-researcher and co-writer and whose death before the age of 55 left such a space in our lives. To our dear Heather Brook.

Contents

Figures

Acknowledgements

This book has taken an age, dogged as it has been by many unexpected and very challenging difficulties—from its lively beginnings in 2014, through various physical, familial and friendship crises and tragedies, the demands of the Covid pandemic, and moving from living in the same city to living at opposite ends of the globe. It has not been easy, and yet there have also been delightful highlights which have come from working together (once again) and from the ongoing love, trust and support we derive from each other.

Chris thanks her great friend Mary for her clear-headed co-writing skills, along with her remarkable good temper and dry wit in the face of challenging demands. Chris also wishes to acknowledge the much cherished ongoing support of her colleagues at the University of Adelaide, located in the Department of Politics and International Relations, particularly Carol Bacchi, Carol Johnson and Tiziana Torresi, and in The Fay Gale Centre for Research on Gender, especially Pam Papadelos and Megan Warin. She is very grateful for the heartwarming collegiality she has experienced in working with her compatriots in Sweden—that is, Ulf Mellström, Andreas Henriksson, Wibke Straube and Lucas Gottzén.

Mary would like to thank Chris for all the good ideas she generously shares and for all the laughter. She is also very grateful to her Sociology colleagues at the University of Edinburgh who have shown true collegiality in their support for her when she was ill and their continued good humour in trying times. Thank you also to her colleagues in the sociology of emotions and especially Åsa Wettergren and Nathan Manning, who co-edit the journal *Emotions and Society* with her. Thanks also to Emily Briggs and Lakshita Joshi at Routledge who have been very patient with us during our trials and tribulations. Mary looks forward to a cup of tea and a catch-up with Emily at some future conference. Mary also wants to thank the friends and family who have provided the best kind of support, with exactly the right mix of sympathy and bad jokes.

We would like to thank Caroline Wamala Larsson and Katherine Harrison in Sweden for their work and enjoyable exchanges with us in our earlier collaborative research study.

And, finally, we both wish to salute our patient, affable and encouraging partners, who offered hugs, conversation, wine and other pleasures to inspire us onwards. Partners are frequently, as in our cases, the unsung heroes of research publications. Dear Brent and John, we throw our hats in the air in appreciation.

Chapter 1

Introduction

> She's had fun dates and boring dates, been pursued and ghosted, charmed and cheated on, fallen in love and had her heart broken, and drunk way more wine on a school night than anyone rightly should. She's met nice guys, dull guys, guys who think they're God's gift, inept guys, hopeless wastes of time, some who didn't look at all like their photos and some who were more attractive in real life, some short, some tall, some super lovely and some total fucking arseholes. And yet, in all that time she has yet to meet Mr Right. So she carries on, war-torn and battle-scarred, living to date another day.
>
> *(Lucy, UK, individual blog, no date)*

Introduction

Internet dating is a relatively novel and increasingly popular practice, defined by Barraket and Henry-Waring as a 'purposeful form of meeting new people through specifically designed internet sites' (2008: 149). In the wake of its growing usage, internet dating has become entangled in wider concerns regarding sexual and emotional marketisation and misuse of personal data by internet companies, as well as associated with deceit, fraud, harassment, abuse and rape (see for example, AnKee and Yazdanifard 2015; Bogolyubova et al. 2018; Rosenbloom 2011). Yet, simultaneously it has also been a site of hope, pleasure, joy and of course desire, romance and love. The characterisation of internet dating outlined by Barraket and Henry-Waring, tendering a careful and blandly descriptive mention of 'meeting new people', hardly seems to cover its rather more complicated and conflicted history, let alone its significance in contemporary societies. We aim in this book to flesh out its tensions, import and implications.

Internet dating deserves to be treated as a distinct phenomenon. However, it is not entirely new (see further discussion in Chapter 5). Commercial dating sites emerged soon after the internet was established in the 1990s. Although the technology is different and there are many sites and apps which each work differently, the basic principles are very similar to previous conduits for connection. Personal advertisements in newspapers and long-established forms of matchmaking all provided ways to find sexual and life partners (Barraket and Henry-Waring 2008; Hardey 2002; Whitty et al. 2007). People across the globe of all ages, gender orientations and sexual preferences use internet dating and some engagement with it has become more and more common (Barraket and Henry-Waring 2008: 150; Couch and Liamputtong 2008; Hillier and Harrison 2007; Jamieson and Simpson 2013: 64–67; Smith and Duggan 2013; Statista 2020a; Whitty 2008a). The growing commonality of the experience and its significance in terms of social life and the development of intimacy provides a strong, ever more compelling rationale for giving detailed attention to examining online dating.

In this context, the central question put forward by this book is, *'Does heterosexual internet dating as a new technology of intimacy offer any opportunities for enhancing social equality, specifically for moving beyond the constraints of gendered heteronormativity and advancing gender equality?'* As Finkel et al. (2012: 3) note, '[o]nline dating sites frequently claim that they have fundamentally altered the dating landscape for the better.' We ask, *'Is this the case for heterosexual gender relations?'* To determine what internet dating brings to social change and, in particular, what it brings to heterosexual gender relations, it is necessary to outline what we understand as normative and what we understand as progressive non-normative possibilities in order to clarify what and how social change occurs.

Internet dating and heterosexuality

We deliberately focus upon heterosexual accounts of internet dating because of our existing theoretical interests in exploring signs of social change within 'the realm of the dominant' (Beasley et al. 2012: 84). Here, we draw upon our previous publications with regard to our understanding of heterosexuality and opportunities for change. We readily acknowledge that sexual relations are not power-free spaces for individual pleasure-seeking. Furthermore, heterosexuality is not merely one option in a fluid range of equally weighted choices. Indeed, as feminist and critical sexuality scholars, we question its dominance and the constraints it imposes. However, relations of domination cannot describe heterosexuality in its entirety. We explore a critical alternative perspective in which heterosexuality is not viewed as an undifferentiated and unchanging monolith and, unlike the common trope in feminist and critical sexuality studies, is not always to be understood as 'nasty, boring and normative' (Beasley et al. 2012 passim).

In short, we reject widely accepted views about heterosexuality as inevitably heteronormative, a term which draws together a conception of heterosexuality as *the* natural, normal and best form of sexuality with a notion of an abiding gender binary in which there are only two forms of selfhood which are different and necessarily complementary (see Warner 1999). In other words, 'heteronormativity' refers to the myriad ways in which social norms of gender and sexuality work together to make heterosexuality, heterosexual power relations, gender polarity and gendered power relations seem given, unquestionable and innate (Beasley and Brook 2019: 14; Beasley et al. 2012: 3–5, 12, 22–23). However, we are conscious that when the term is used, the axis of gender is sometimes less evident and hence throughout the book we have referred to 'gendered heteronormativity' to render its interwoven meaning in relation to gender, as well as sexuality, more visible.

To suggest that heterosexuality is not always equivalent to gendered heteronormativity challenges the hegemonic coherence of heteronormativity and thus resists constituting change as only available at the social margins. Rather, we draw attention to the 'micro-politics' of change in the domain of the mainstream and everyday, such that change cannot be presumed to be top-down or linear but rather is viewed as diverse, highly uneven and potentially unpredictable. Our overall approach is a socio-political one but that approach is shaped by concern with what we term micro-politics: this term registers a broadened conception of 'the political', of power relations. Such an enlarged conception enables re-consideration of the familiar and commonplace and invites consideration of inconsistency and plurality. It includes not only rational cognitive activities but embodied, emotional, libidinal aspects of conduct. Furthermore, micro-politics provides a different perspective on the scope and scale of 'the political', providing a bridging terminology between localised, private innovations and larger-scale, organised movements and institutional developments usually connected with more public terms like 'dissent' and 'protest'. We consider an approach which expands understandings of the political as especially appropriate to examination of internet dating in that internet dating involves everything from intimate exchanges between individual women and men through to the involvement of institutions like the family or religions. Our approach considers social change in a framing which includes small-scale private innovations, that may be deeply personal but are also linked to macro social assemblages, as we explore the complexity of heterosexuality. Moreover, this approach is marked by our willingness to investigate meanings and practices of heterosexuality in heterodoxical ways. In other words, we not only intend to counter the comparatively limited and primarily negative existing scholarship on heterosexuality, we do so in ways that attend to heterosexuality's intricacies enabling teasing out of unconventional directions—that is, of divergent, transgressive, subversive, dissident and pleasurable elements. These non-normative elements are not necessarily dramatic: we do not so much aim to explode the notion of the immutable monolith of heterosexuality as investigate its marbled contours and fissures (Beasley 2015; Beasley 2011; Beasley et al. 2012: 84). In this volume, heterosexuality's complexities come to light through the lens of internet dating.

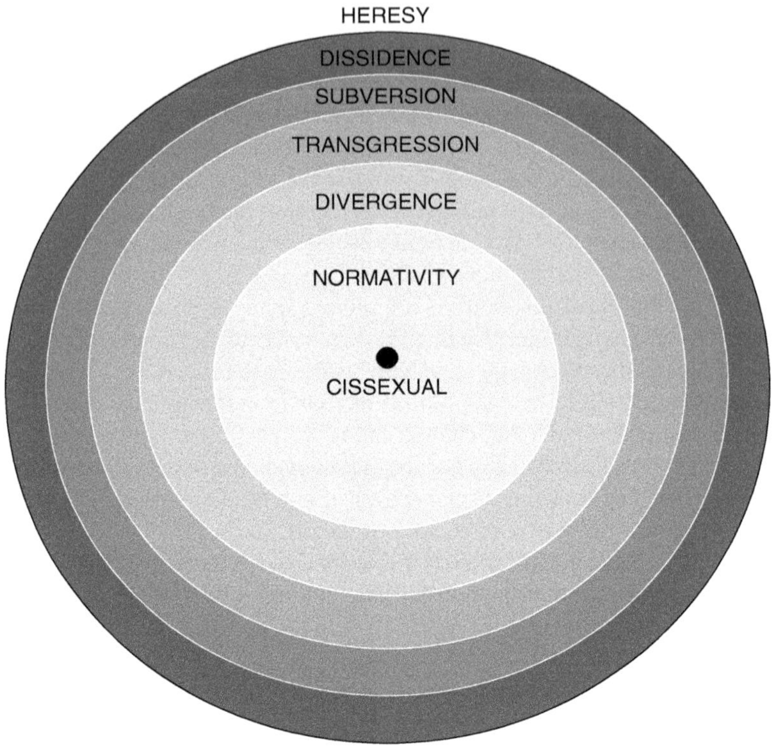

Figure 1.1 Heterosexuality and heterodoxy—from heteronormative to heretical. A colour version of this figure is downloadable from: www.routledge.com/9780415720694.

Our alternative view of heterosexuality enables a frame of reference in which it is considered as a diverse range of practices, providing a means to investigate departures from gendered heteronormativity, as portrayed in Figure 1.1.

In the centre is 'cissexuality',[1] where heterosexuality does match with heteronormativity, where sexed bodies, gender and sexual practices largely conform to hegemonic definitions of the normal (Harrison 2013: 12–13). Contained within the wider inner circle might be slightly less hegemonic, but still satisfactory options in relation to normative requirements. Beyond this the figure displays the non-normative possibilities for heterosexuality, from divergent to 'heretical'. *Divergence* describes practices linked to the norm but departing, to a limited degree, its boundaries. For example, normatively, and often legally, marriage has been expected and indeed required to involve sexual consummation as part of what is supposed to be a close link to reproduction. However, there are instances which breach this link such as the case of a marriage being judged not to have been consummated and hence capable of annulment, even though the wife became pregnant from her husband ejaculating between her legs (Beasley et al. 2015: 687). Such an example, along with many others, to some extent diverge from normative expectations of heterosexual performance, showing that even at the institutionalised heart of heterosexuality, practices might not quite fit the ideal. Further 'out' from divergent practices arise somewhat more challenging possibilities. *Transgression* from heteronormativity involves temporary and not usually deliberate departure. A couple, for example, may occasionally enjoy 'swinging' relationships where they have sexual relations with other partners, without this challenging their otherwise conventionally heterosexual relationship in which they privilege emotional connection to each other (Visser and McDonald 2007).

More conscious and continuing undermining of heterosexual norms is described as *subversion*. Subversion may not always be radical but might upset heteronorms in quiet ways. An example here might be ongoing non-cohabitation amongst established, committed heterosexual couples (Beasley et al. 2015: 689–690). Subversion blurs into *dissidence*, which is more radical and more intentional in its departure from the norm. Dissident forms of heterosexuality might include those that problematise homosexual/heterosexual boundaries such as some bisexual practices, or those like celibacy that entail an alternative (even if possibly temporary) direction (Donnelly and Burgess 2008). Last, we come to *heresy*. Heretical forms in some way or another involve a refusal of heterosexuality. These forms include homosexuality, non-binary challenges to the gender distinction which disturb the very basis of heterosexuality, and asexuality which questions the centrality of sexual activity per se to personal identity and social life (Scott et al. 2014).

Where the boundaries between each 'level' of the circle lie is unclear and questions about how or why people might travel out or back, or 'swivel' between these different levels await further research, but it is clear that this model assists in the consideration of gender and sexuality in more fluid ways, both within heterosexual practices and across the permeable boundaries which supposedly demarcate them from, for example, practices associated with same-sex, trans and non-binary experiences.

This elaboration of the non-normative in heterosexuality provides a framework for analysis of internet dating as a potentially exciting form of social connection and intimacy amongst heterosexuals and one that may promote progressive social change. The framework highlights how heterosexuality online may be understood in more nuanced ways and is used throughout this book to consider our central question regarding the potential of internet dating in terms of advancing social innovation (Beasley et al. 2015).

Researching internet dating

Existing studies of internet dating tend to be descriptive and typically give little attention to progressive change towards enhanced social equality.[2] On this basis, further exploration of experiences of internet dating is warranted. This book not only focuses on possibilities for social change, specifically in relation to troubling gendered heteronormativity and advancing gender equality, but additionally foregrounds accounts by internet daters and our analysis of their experiences. Despite the increasingly extensive body of analyses and research studies that provide conceptual and empirical commentaries upon heterosexual internet dating, they are not all-encompassing. Current scholarship is largely 'Western', but it is even more clearly disproportionately North American (Cardona 2019; Kisilevich et al. 2012; Luo 2017; Rosenfeld 2018). We contribute something new to this body of work by noting commentaries and research arising from a range of social contexts including 'non-Western' sources. All the same, although the book has an international flavour, our discussions about internet dating are primarily derived from 'Western' and English-speaking sources and sites. In particular, with the exception of Chapter 7, our online accounts are from internet daters based in the UK and Australia.

The book has some further noteworthy restrictions. We have concentrated upon our central concern with intersections between gender and sexuality. Intersections between race/ethnicity/culture, gender and sexuality are given some attention in Chapters 4, 5 and 7, and Chapter 6 specifically deals with intersections between age, gender and sexuality. However, intersections between class and gender/sexuality are only occasionally noted (see for example, Chapter 3) and largely fall outside the scope of this book. It is also worth noting here the partisan weighting of existing sources about dating—as well as associated terms like intimacy, romance and desire—arising from psychological and health-oriented literature, as well as the tendency to divide off dating from sexology and sexual activities (see for example, Döring 2009; Eleuteri et al. 2014). By contrast, while we do reference the psychological and health literature, our perspective is primarily a socio-political one focussing on potential for social change in relation to gender and sexuality. Moreover, we bring gender, sexuality and dating into conversation under the rubric of intimate sociality.

Here, it becomes necessary to clarify our understanding of intimacy. While sexuality may be viewed as within the continuum of intimacy, it is also useful to distinguish them. The term sexuality encompasses not merely sexual acts, practices and experiences but also a conglomerate of institutions, identities, social assumptions and customs, as well as resources and labour (Beasley et al. 2012; Jackson 1999). By comparison, intimacy for us is usually employed to refer to a more diffuse arena than sexuality, concerned with a sense of close, embodied and particularised personal connection (Budgeon 2008; Henriksson 2014; Jamieson 1999; Roseneil and Budgeon 2004), as well as physical affectionate touch (Debrot et al. 2017; Debrot et al. 2013). This is typically not only about the individuals that experience it but also embedded in relationalities beyond it (Smart 2007). In attending to the micro-politics of internet dating we conceive sexuality and intimacy as linked, though not reducible to each other.

In this context, we note Lauren Berlant's conceptualisation of intimacy as a

> narrative about something shared, a story about both oneself and others that will turn out in a particular way. Usually, this story is set within zones

> of familiarity and comfort: friendship, the couple, and the family form, animated by expressive and emancipating kinds of love.
>
> (Berlant 2000: 1)

Our concern with intimacy in this book also examines a well-known shared narrative—about (heterosexual) dating—but departs from Berlant's stress on the familiar and comforting. Rather, we wish to examine the ways in which a narrative of dating may activate zones beyond the known and secure and animate a range of possibly unexpected connections. Rather than a term that is tied to what is felt to be cosy, we focus on the transformative potential of internet dating. Since internet dating has become an ever more significant method for developing intimate relationships, it is important to investigate its possibilities in terms of considering links between intimacy and social change which potentially disturb theorising about what social change might be and how it might appear. This book is intended, in other words, to provide a picture of the *socio-political possibilities of internet dating, as a technology of intimacy, in a primarily 'Western' frame.*

To flesh out this picture, we have employed a variety of sources. Dating websites regularly undertake research and may permit scholars to develop and publicly analyse their data sets, thus providing an important, if partial source (see for example, Lee and Niederle 2011; Ong and Wang 2015; Ortega and Hergovich 2017; RSVP Date of the Nation Report 2015). We make use of both the information which the websites provide and academic analyses of the information. We attend also to social science and other scholarship (which includes methods such as interviews with internet daters), to contributions and conversations about dating such as blog posts, and to journalistic and other commentaries, such as those by 'experts' including psychologists, therapists, sexual counsellors and relationship coaches (for example, Herbenick 2014; Leggatt 2017; Shpancer 2014). Furthermore, as indicated earlier, this book deals with socio-political approaches to intimacy in relation to gender and sexuality, rather than having a primary focus on the internal psychological or health risk literature, and deals with a variety of scholarship concerning relationship formation and sexual attraction (Allen 2003; Asendorpf et al. 2011; Beasley 2017a; Eastwick and Finkel 2008; Fisman et al 2006; Garcia et al. 2012; Gillies 2003; Lenton and Francesconi 2010; Sassler 2010; Sassler et al. 2018).

However, we also draw upon of our own study which has as its core focus the experiences of UK and Australian heterosexual internet daters. We attend to the perhaps surprisingly extensive range of online narratives of their experiences. From these burgeoning sources we selected a sample describing hundreds of dates, appearing in more than 40 blogs, articles or collections of stories which describe the dating experiences of at least 70 different individuals from the early 2000s until 2020. Roughly half of the daters are from the UK and half from Australia. Sometimes the authors who are recording their experiences of internet dating are themselves unsure how many dates they have been on, but some individuals provide accounts of scores of dates over many years, others record just a few and some only share one of their experiences or talk generally about dating online. Some of the accounts appear in blogs, others in online newspaper articles, while some internet dating stories are collected by journalists or shared via discussion forums and other platforms. These sources were purposively selected using internet and blog search tools and by seeking blogs attached to internet dating sites, especially from the most popular mainstream sites. In order to provide empirical examples beyond the usual dominant North American ones, we have attended to UK and Australian heterosexual internet daters as our central concern but have also noted examples from other countries (see Chapter 7). In part this material arises from a collaboration with colleagues who focussed upon Denmark and Uganda. That collaborative research project involved a qualitative content analysis of internet dating sites and examines a sample of 20 profiles in each of the four countries. Beyond this collaborative study, we also undertook a more schematic analysis of geographically and culturally dispersed instances of dating sites and profiles from three further countries in South America, Asia and the Middle East for further comparison.

In both the collaborative and schematic research projects, we particularly concentrated upon dating sites and profiles as a means to examine presentations of self, attitudes and practices. The collaborative project considered five major dating sites and 80 profiles (40 men and 40 women) overall across the four designated countries. The Denmark and Uganda studies also involved posting a question to chat sections or Facebook groups related to some of the dating sites asking if users could identify new or different experiences of internet dating compared with traditional experiences of dating. The schematic analysis from an additional three countries (Argentina, China and Turkey) involved examination of three main dating sites and 30 profiles (15 men and 15 women).

In attending to daters' experiences, we have tried to maintain a distinction between 'found' data and 'made' data (Stanley and Sereva 2019). 'Found' data is not produced by 'researchers' (Stanley and Sereva 2019)—even if we have later made selections from these accounts—but by individuals depicting their own experiences of internet dating to share. 'Made' data is reported/mediated where someone else (often a journalist, professional commentator or researcher) has sought, collected and organised those experiences (Stanley and Sereva 2019), thus shaping their initial articulation as well as their subsequent selection. This 'made' data nevertheless still provides a record of the voices of internet daters. We use these terms because commonly employed distinctions between 'primary' and 'secondary' data are difficult enough to maintain offline (Bishop 2007, Moore 2007) and proved to be not always helpful for categorising the online accounts that constitute our empirical material. Many of the accounts we analyse of the experiences of UK and Australian internet dating are shared in individual blogs, blog posts or media stories written by the internet daters themselves. A rich supply of data came from individual blogs by a few UK daters who recorded their internet dating over a short period involving up to 30 dates, or for over a decade or more, providing accounts of hundreds of dates. A couple of the other blogs were written by various authors on general lifestyle topics, occasionally including accounts of internet dating. One Australian general blog site, which attended to the topic of internet dating, generated an online conversation over two years, offering an abundant source of experiences and debates about its merits. Some blogs specifically about internet dating were associated with commercial dating sites and had a variety of contributors. Additionally, several accounts of internet dating were found in blogs written by 'expert' men concerned with coaching men to find women online.

Apart from blogs, accounts of personal experiences of internet dating regularly appear in online newspapers. We treat as 'found' data those articles written by the person who did the dating. Also useful however, are daters' accounts collected and organised by others and we indicate use of this 'made' (reported/mediated) data by attributing it to the person who gathered the original material. As an example, there are online newspaper articles where journalists, professional counsellors, dating advisors and 'coaches' and academic researchers, among others, have interviewed internet daters and directly quote what they have to say. This material is treated as 'made' rather than 'found' data because it has been shaped and then selected by people other than the daters, and the daters' version of the interview is not available. It is difficult in this circumstance to know how representative the 'made' samples are of daters' experiences. For instance, it is possible that authors might have omitted things that the daters thought important, but we cannot check for this against the daters' version.

Of course, there are always questions that can be raised about any selection of data, including our own. However, where we select from accounts daters have written themselves, it is possible for us to see the entire account as they wrote it. For this reason, when we have referred to accounts 'made' by someone other than the person who had the experience and where we do not have access to the entire account as the person gave it, we signal this distinction. That said, such 'made' accounts constitute an important record of daters' experiences over time in popular culture. For instance, they constitute a form of documentation before academic research provided much in the way of access to their experiences. Moreover, the 'made' accounts do give us insights into some of the aspects of internet dating that are not always apparent in the self-authored 'found' accounts. In particular, they help us say more about women's experiences, as there are fewer detailed self-authored accounts of dates by women in our sample of UK and Australian daters. It is evident that the web can be a hostile place for women sharing their opinions and experiences (Jane 2014). Thus, although we consider it important to note when we are referring to 'made' accounts, by using them alongside 'found' self-authored accounts we are able to represent a wider range of women's experiences (see for example, Albo 2017 and Chapter 5).

We also cannot know exactly how representative the 'found' or 'made' accounts we have collected in our study are of the general experiences of internet daters or even just those in the UK and Australia. First of all, these experiences are likely to be very different in different national/cultural locations and at different times and for different age cohorts. For example, internet daters in China are apparently much more likely to be seeking marriage than those in the UK and Australia (Li and Lipscombe 2017; YouGov 2017a). Similarly, while internet dating is increasingly socially accepted and widely used in many countries, thus far it remains comparatively less accepted in a range of countries including China (Smith and Anderson 2016; see also Chapter 7). Cultural differences can be very stark, as can be seen in the uneven acceptance and take-up of online dating globally in countries like Turkey compared with the four countries in our collaborative research study. This is exemplified by the very much lower rates of online dating in Japan compared with neighbouring China (Paisley 2018).

Second, even within the UK and Australia, the accounts we have employed in our study sample necessarily arise from those who are willing to share their experiences. This may be because they have been 'successful' in online dating, because they want to help other people by sharing negative experiences, because they are being used as a form of advertising by commercial mainstream dating sites, because they are wanting to establish themselves as writers with well-read blogs and think this a topic likely to attract attention, or for other less evident reasons. Whatever their reasons, the blogs we make use of are likely to be written by certain kinds of people, as generally the blogosphere is populated by well-educated, young and middle-class people who are usually white (Hookway 2008; Lenhart and Fox 2006; Prescott et al. 2015; Snee 2013: 147–149). This limitation is very likely to also apply to all the accounts we analyse, as those who have 'made' accounts do not seem to have sought a diverse range of interviewees.

As we have noted, scholarship on internet dating has concentrated upon younger daters. Our sample has its own limits, being drawn from a pool of people who all have some digital competency and internet access because those are necessary requirements for sharing experiences of internet dating online, and indeed for doing internet dating. Yet, there are many internet daters who do not write accounts of their experiences and they might have rather different stories to tell of how gender, sexuality, race/ethnicity or age impact on their dating experiences. Although other scholars have used interview methods to enable access to a variety of dating experiences (Anderson 2012; Hakim 2012), with some interviews undertaken via online chat (for example, Couch and Liamputtong 2008, Frohlick and Migliardi 2011; Hillier and Harrison 2007), we decided to foreground online accounts by daters themselves. Journalists, professional commentators writing in the media, academic researchers and others who interview daters will conventionally obtain agreement from interviewees that the account they publish is acceptable to the interviewees, but nevertheless are engaged in the shaping of daters' accounts at the point these accounts are articulated. By contrast, while we make no claim to our sample being statistically representative, we considered it could provide meaningful qualitative clues precisely because many of the online accounts are produced without the prompting of interviewers, and thus can tell us what internet daters feel is important to relate about their experiences (Hookway 2008; Snee 2013: 147–149). These '[n]atively digital data objects . . . provide a window into social practices and identities that take place when people are not consciously aware that they are being surveyed, interviewed or otherwise canvassed for their opinion' (Lupton 2015: 44).

The authors of online dating commentaries are providing these accounts for an audience and wanting to tell a 'good story'. They are likely to be selective in what they share and to edit their experiences into familiar narrative forms and, in particular, into tropes of romance. Indeed, the type and shape of these common narratives about internet dating forms the topic of one chapter. Otherwise however, we thematically analyse the online accounts in order to illustrate the kinds of possibilities that might exist for internet dating to generate moves beyond the gendered heteronorm amongst these relatively privileged heterosexual groups. The extent to which these possibilities might be apparent within wider populations must await further research.

The ethics of using online material for research are complex, even though all of the material drawn on in this study appears in the public realm (Baker and Whitty 2008; Beer and Burrows 2007; Hookway 2008; Sugiura et al. 2016; Townsend and Wallace 2016). There are debates about whether online communities and forums are public or private spaces and whether consent is necessary. We tend toward the view that the data we have used is publicly available, not confidential, and that it was not necessary to advise the individuals of our research (Lupton 2015: 63). However, we have taken care to ensure that no harm is likely to come to any of those whose accounts we have cited.

We were granted ethics approval for the research used in this book by both the University of Adelaide and the University of Edinburgh, but the ethics of online research is a rapidly shifting field and worthy of at least brief comment here. Those sharing their dating experiences in blogs are deemed to be 'publishing' their writing but, rather than treat them like published sources, we treat them more like interview transcripts, giving them pseudonyms if they appear to be using their real name, and not associating the quotes from posts with links to the original blogs from which the quotes have been derived. We have similarly anonymised others sharing their accounts in publicly accessible forums, even though they may be considered to have consented to strangers reading what they have written. All who share these online accounts may not be expecting researchers to analyse them. In addition, the accounts detail dates with other people. Some of the writers, especially the bloggers, appear to have gained the consent of those they dated when publishing a story referring to their dates. The writers and the other accounts of internet dating are almost always careful not to provide details that might identify the people they dated. Whatever the case, the 'public' nature of these online accounts is open to question and thus for all accounts we provide a further level of anonymity beyond pseudonyms, by ensuring that there are no identifying details in the extracts from internet daters we present here.

The quotes from our 'found' data are labelled without referencing the source they come from, and presented in italics according to the formula of a pseudonym we have chosen for the author, which indicates the apparent gender of the participant, country, the type of source and year of publication (for example: *Rose, Australia, dating blog, 2013*). The pseudonyms are designed to help readers keep track of the different daters to whom we refer. The format for citing the 'found' data will maintain some anonymity but the accounts are searchable and therefore can provide a set of retrievable data, allowing readers—if they wish to look further—to evaluate our interpretation of the accounts as a whole (Stanley and Wise 2006). We consider that this is a compromise between granting an anonymity that many of those sharing their experiences online may not seek and exercising some caution regarding those whose stories are being shared and used in ways they are unaware of (see Moore 2012). It also, we hope, makes it easier for the reader to follow the stories presented.

We follow a similar format in relation to internet daters' accounts presented in the 'made' data. Here, however, we make use of a pseudonym for the dater (sometimes one provided by the author) and also provide attribution to the author responsible (for example, *Gina, Australia*—cited in Nguyen 2018). These citations can be followed to our reference list. This format for 'made' data enables us to give due recognition to the dater, while also acknowledging the person who sought and selected the dater's views and authored the work, thus giving context.

Organisation of the book

Having established a theoretical and methodological framework for our investigation of internet dating in this Introduction, we turn in the second chapter to the experiences of internet daters and begin by establishing where the boundaries of innovations in internet dating might lie. We do this by examining a common narrative told about internet dating experiences, which we call the 'nutter' narrative. In exploring this narrative, we can discern what some of the established norms around gendered heterosexual interactions in dating might be by employing the time-honoured technique of seeing what reactions ensue when norms are breached (Garfinkel 1964).

The third chapter sketches the terrain of new norms emerging within internet dating and examines to what extent they trouble heterosexual gender hierarchies. The relational and participatory formation of norms relating to internet dating moves as these hierarchies alter. We examine to what extent contestations around etiquette demonstrate the kinds of reflexivity required in navigating social innovation.

In Chapter 4 we explore how this reflexivity is emotional, enabling internet daters to navigate towards more heterodox (non-normative) heterosexualities in ways that are exciting, not just fearful. We explore these heterodoxies, from divergent to transgressive and occasionally subversive, ways of doing heterosexuality and counter the usual emphasis on dating as a rationalised activity like shopping; instead exploring a range of emotional and embodied practices that occur in navigating its pleasures and pitfalls. Chapter 5 further develops our exploration of to what extent internet dating constrains gendered heteronormativity and encourages gender equality by focusing on what is arguably its raison d'être—chemistry. Accounts of internet dating represent chemistry as essential, but what it means is not always clear. We investigate this term and the ambiguities around its use.

Having spent the first part of the book examining heterodox possibilities in the realm of the dominant by elaborating on key aspects of heterosexual internet dating, in two chapters (6 and 7) before the concluding one we assess how these possibilities might differ with age and with geographical location. This does not capture all the varying intersections between gender, sexuality and other axes of power. The central aim, however, is to consider the relationship between social innovation, gender and sexuality as they are done differently at different stages in the life course and in different locations. Thus, Chapter 6 turns to examining the experience of older internet daters to find that they experience heterodox innovations which are politically significant.

Drawing on a collaborative exploration (Beasley et al. 2017), in the seventh chapter we explore new modes of developing intimate relationships arising out of the growing use of internet dating sites globally and in attending to geographical difference we engage with issues of race/ethnicity/culture. We consider potential contributions of internet dating to unsettling gendered heteronormativity in Uganda and Denmark (in which our collaborators have lived or now work), as well as in the UK and Australia (where we currently live and work). As noted earlier, the method used in this collaboration differed slightly from that employed in other chapters, being principally based on analysis of 80 dater profiles from four countries (UK, Denmark, Uganda and Australia). This small collaborative study, in combination with reference to culturally dispersed illustrative instances from three further countries (Argentina, China and Turkey), allows us to examine the potential for our theoretical model of heterodoxy to be extended beyond 'Western' contexts. The concluding chapter draws together all of these materials to provide a condensed account of the argument and content of the chapters, as well as summarising the surprising array of heterodoxical innovations we have found.

This book offers a significant departure from the main directions in analysis and scholarship dealing with internet dating. There are a range of ways in which our approach differs in important respects from prevailing perspectives and provides novel directions for apprehending this burgeoning technology of intimacy. Our investigation of internet dating provides a theoretically grounded consideration of social innovations arising within internet dating, especially as they relate to social equality in regard to gendered (hetero)sexuality. Our model of the heterodoxies within the dominant realm of heterosexuality, counters thinking which assumes that heterosexuality is unexciting, unrelentingly normative and unremittingly an arena of hierarchical relations of dominance and subjugation (see also Beasley et al. 2012). We here apply that theoretical model to internet dating, while maintaining an awareness of the limitations of innovations and the policing of attempts to step outside of gendered heteronormative constraints on behaviour (see Bean 2011).

Additionally, our use of narrative as well as thematic analysis of internet daters' experiences is unusual in internet dating research and assists us in illustrating where limits to and opportunities for change might arise. By looking at a range of accounts by daters we also see where new norms are emerging. Those insights include examining the importance of emotional reflexivity and embodiment in internet dating and giving attention to the centrality of chemistry, arenas which are neglected within most research in the field. We also resist the exclusionary focus upon and presumptions associated with the dominance of concern with young daters and instead give significance to the experiences of older internet daters. Similarly, we reject the predominant concern in the popular and scholarly literature with North American data, social mores and daters. Instead we extend our central attention to UK and Australian daters to also include daters in other international contexts including consideration of certain national instances in Scandinavia, Africa, South America, Asia and the Middle East. Most of these instances are typically given little recognition and rarely surveyed by empirical research. Through analysis of such varied experiences we are able to question the usual association of social innovation with young people, and arguably with the Western 'developed' world. We thus present internet dating as not always experienced through marketisation or fear of its risks, nor as rational cognitive and disembodied, but as fun, fleshy and full of intriguing and unexpected opportunities for social change–even if these opportunities are not without limits.

Notes

1 This terminology is still emerging and there are several meanings attached to it. Nevertheless, it is typically located as the antonym of 'transsexual' and in our usage combines 'cisgender' (alignment of sex designated at birth with gender identification) and 'straight'—that is, we use it as a shorthand for a relatively clear-cut alignment with gendered heteronormative heterosexuality. See also Urban Dictionary, definition of cissexual and Oxford English Dictionary Online, definition of cisgender.

2 There are exceptions to this assessment, notably McWilliams and Barrett (2014) and Bean (2011). However, more positive considerations of change may also be aligned with questioning feminist/gender/queer critiques of modern sociality. See for example, C. Hakim (2012) *The New Rules*. We aim, by comparison, to enable analysis of possibilities for change precisely in the context of critical gender/sexuality studies.

Chapter 2

Nutter narratives and the boundaries of heterosexual gender norms online

There are joys and sorrows in internet dating. And the joys may come as much from meeting a range of interesting people as from finding a soulmate. We look at how some of the many intriguing stories to be told about the crazy conversations and the ups and downs of looking for love online can act like force fields that remind people of the shocks they may receive if they go beyond the norm. We begin our account of internet dating in this way because if we are to chart the innovations that it enables, it is important to have some sense of their limits. As the book unfolds we will journey out through the realms of divergence, transgression and subversion, with occasional ventures into dissidence, as we see how internet dating provides opportunities to do relationships differently. However, our data does not, for the most part, depict markedly radical directions. Even moments of dissidence tend to be unassuming and do not usually offer a provocative challenge to heterosexual gender relations—though, as will be pointed out in later chapters, unassuming possibilities may well facilitate significant social refashionings. Here we consider why there is not more heterodoxy, let alone queering, evident in the experiences of internet dating we examined.[1] This chapter therefore details examples of what Ahmed (2006: 92–96) describes as 'straightening' or 'stopping', whereby heterodoxy is made sense of in ways that reincorporate it within gendered heteronormative views of the world. We examine how heterosexuals' attempts to trouble gender norms are made sense of in ways that reincorporate it within normative world views. This 'making sense' can be seen at work in popular narrative forms used by internet daters.

Prominent in the blogs we examine in our research sample are stories of dates gone wrong, of meeting a 'nutter' or crazy person.[2] We focus on these in this opening chapter because of what they reveal about how boundaries around gender norms are policed in internet dating. A number of other narratives are dealt with in other chapters, such as narratives of desire and of finding—or more commonly not finding—chemistry (see Chapter 5), but nutter narratives appear frequently. The predominance of this story in the blogs should not be taken as evidence of it being one of the most common experiences. To the contrary, the majority of posts suggest that what is more common are mundane experiences of meeting, having a nice time, but finding there is 'no spark' or being in other ways disappointed. However, the nutter story is so prominent as a narrative precisely because it is unusual and makes a good story. It is likely, as we shall see, to have elements of comedy about it and to be told more as a kitsch kind of gothic horror or adventure story. One blog that started because of an encounter with a 'nutter' even has an illustration of red blood trickling down the side of the webpage. Yet, the regular occurrence of this narrative has much to tell us about how people navigate gendered heteronorms as they manage their internet dating relationships. We begin by outlining the nutter narrative and then proceed to analyse each stage. The narrative helps illuminate the journey of internet daters from pleasant, untroubling initial contact and meeting, to the emergence of doubts, to pushing on despite those doubts. It helps explain when and why various parties might feel cautious or retreat from contact. Nutter narratives are clearly gendered and we suggest that while both the men's and the women's narratives might reinforce various gendered heteronorms and expectations, they also tell stories which give a glimpse of new ways women might exercise more control over whether, when and how relationships proceed. Nutter narratives tell us how these various efforts to do gender differently are limited—that is, in Ahmed's terminology, are 'straightened' up, or even 'stopped'.

Internet dating narratives

People like to tell stories about internet dating. Whenever we have presented ideas from this book our audiences have shared with us some of their own stories. These stories are often of meeting the love of their life online, or the weird and not so wonderful dates they have met through internet dating sites. The blogs and other accounts on which this book draws are further testimony to the way in which people want to make sense of, and share, their experiences through narrating stories. Much work on narrative neglects to define it and in any case there are significant debates about its meaning, but in this volume we take it to refer to 'stories that are linked together in life' and a story as an account of what has happened to people, 'which has a beginning, a middle and end, although not necessarily in this order' (Stanley 2008: 436–437, see also Flynn 2019; Squire et al. 2013).

The way in which internet dating narratives are developed has a lot to tell us about what is important to people and how experiences are made sense of. As Jerome Bruner (1987: 12) argues, narrative is the only way we have of describing 'lived time'. Thus narratives impose some order on what has happened. They are however, stories of experience and not actual experience (Squires 2013). In fact, narratives make worlds and construct gendered selves (Bruner 1987; Butler 2004). They do so because they are not simply individual stories but 'highly susceptible to cultural, interpersonal and linguistic influences'

(Bruner 1987: 14). Narratives about romance offer a mode of storyline, often influenced by and interpreted through well-worn fictional forms as well as stories found in everyday social circles. Romance narratives, whether outlining 'success' or 'failure' in love, largely follow a genre format with the achievement of (gendered) selfhood at their core, and enabling the possibility of fantasy, recognition and nurture (see for example, Juhasz 1998: 65). The point here is that narratives about internet dating do not arise in a vacuum but involve contributions to a long established format, even if that format has many variations—a format that is highly responsive to social norms about intimacy and interconnection.

In this context, we highlight the recognisable similarities in the stories told, the almost formulaic nature of many of the accounts of internet dating. People experience internet dating as a series of events, usually in sequence. The use of narratives to make sense of these experiences is striking in the sample we examined. It may not seem at all surprising that blogs and other online accounts contain narratives which are are efforts to describe people's lives and in particular romance in people's lives. Blogs in particular are usually designed to try and attract an audience and thus it is important to tell a good story (Eisenlauer and Hoffmann, 2010: 79). The blog is also close kin with the diary, a highly personalised form of autobiographical writing less intended for (immediate) public consumption (Arioso 2010). Whether they appear in blogs, online articles or other online sources, we have selected material that tells stories about the experiences of internet dating and that foreground the narrative voices of internet daters. Thematic analysis alone did not seem adequate to do justice to the similarities and consistency in structure of these stories and thus a small-scale analysis of the narratives was done. These narratives were analysed by considering what meanings the tellers want the audience to take, by noting the connections the stories make between past, present and future and by examining their interactional context.

Through the nutter narratives we see to what extent selves and gender subject positions are being normatively produced and practised in interaction with other daters and with the imagined audiences reading the accounts (Phoenix 2013). These constructions of selves and subjectivities involve efforts which the narrators make to render themselves as intelligible 'women' or 'men', in the ways that Butler (2004) has suggested. We sought stories that were 'emblematic' (Phoenix 2013) of the challenges of navigating gendered heteronormativity in these online practices. Often normative expectations are best revealed when norms are breached, when things do not go entirely as expected or when what are perceived as discrediting stigmas are revealed (Garfinkel 1964; Goffman 1968). This is what happens in the nutter narratives, which reveal how boundaries are maintained around heterosexual gender norms. Internet dating may allow opportunities for diverging from, transgressing or subverting such norms, but there may be consequences for going 'too far'.

Heterosexual gender norms online

Most of the evidence suggests that the internet contributes to the reproduction of gender norms and inequalities, despite some opportunities for changes resulting in more equality (Brickell 2012: 38; Jamieson 2013). This is not entirely surprising because sustaining relationships online usually means moving into offline communication (Baym 2015). In fact, 'internet dating is typically used with the intention of quickly progressing to face-to-face meeting in search of the right "chemistry" to sustain a co-present relationship' (Jamieson 2013: 29; see also Chapter 5). Yet, it is important to recognise the opportunities internet dating provides 'for flirting or getting to know potential sexual partners which are far removed from patriarchal dating systems in which men choose women and women wait to be chosen' (Jamieson 2013: 22). The elaboration of those opportunities requires more understanding of power relations online.

The general lack of attention to the operation of power relations with regard to gender and sexuality on the internet (Brickell 2012) leaves the exact nature of the pitfalls and possibilities for innovations in heterosexual gender relations largely unexamined. Brickell's analysis considers power as constitutive, regulatory and unequal and he concludes that power relations on the internet are 'multi-directional'. This is not as illuminating as it could be, but the application of Foucault's models of discursive power, blended with Althussarian focus on interpellation and feminist analysis of inequalities can be useful in seeing how the internet makes certain kinds of gendered subjects. The details of this need teasing out in relation to internet dating. Where explored, the affordances of the technology may be examined in terms of how, for example, pre-set grids ('I am looking for . . .') constitute internet daters as gendered subjects and how internet dating sites entail regulatory power-producing forms of pleasure (Brickell 2012: 31–39). Although the internet may allow for more transformative gendered sexualities, much of the work acknowledging that possibility does not explore to what extent such transformations occur, under what conditions, and what they may look like. It is here that our work can contribute. We consider how narratives, in this instance the nutter narrative, draw on discourses about heterosexual gender relations and seek to produce conforming gendered subjects. However, these narratives also tell us about spaces for resisting what Gagnon and Simon (2005/1973) describe as the usual 'sexual scripts' governing gendered heterosexual relationships. The stories also, at least to some degree, speak of crossing beyond the bounds of the normative. While use of this terminology of scripts has been disputed (Beasley 2012; Beres 2014; Longmore 1998: 54–56), nevertheless for our purposes it draws attention to the social constitution of intimacy rather than viewing personal relationships as driven by biology or individual psychology, strategising or values. In order to comprehend this sense-making we need to outline what the nutter narrative is, before critically dissecting it.

The nutter narrative

A blog post explicitly entitled '[Arlene] the internet dating nutter' provides a succinct and classic example of the nutter narrative that we use as a basis for our analysis. It is written by a divorced man from the UK who blogs about two initial internet dating experiences. In relation to this particularly apposite blog we have provided pseudonyms for both the narrator ('Peter') and his date ('Arlene') in order to protect her anonymity. We will also use another of Peter's stories, one about his first experience of going on a date with someone met on the internet. This story of a first online experience is somewhat ambiguous but does contain elements of the 'nutter' narrative. It involves him going to Moscow to meet a woman (we will call her 'Natasha') he has found on *Match.com* and gradually concerns arise about her truthfulness. We mention additional examples of nutter narratives to further flesh out reiterated patterns. In order to illustrate the format of the nutter narrative, we will firstly set out the exemplary nutter narrative as it is told of Arlene.

After some preamble stating that Peter met Arlene via a particular dating site and a digression about the city where Arlene is from, the narrative begins with his story about actually meeting her.

> We met for the first time and went to a restaurant near Newton le Willows. It was a very nice evening. Then we had a pleasant enough evening in Manchester, which lead to a kiss. She was very good at kissing. She then started to unravel what she had told me and started to put things in to two camps. Fiction and non-fiction. She was blonde with blue eyes. She wasn't 39 (any more) she was 42. She wasn't a beautician, she was on a course learning about painting nails. She hadn't been single for a long time, she'd been single for about two months after her boyfriend, who used to beat her up and anyone who looked at her too, finally went too far, leaving her with a mountain of debt.
>
> I was starting to get cold feet. We had another lunchtime date in Liverpool, where we happened to bump in to cousins of hers. I did wonder if the meeting was manufactured, but didn't ask. A couple of days later I was in a pub with a couple of friends. I had just taken a mouthful of drink, when I received a text. It was a short text, all it said was 'I love you'.Unfortunately for my friends, I couldn't keep my drink in my mouth and it ended up all over them! This was still early days for me, so naively I suggested we meet up again. We met at the Albert Dock in Liverpool. I explained to her that things were moving way too fast for me. I kissed her on the cheek and left. Nothing happened for a couple of hours. Then I received a text. It said 'Are you sure you've made the right decision'. I didn't reply. Then a couple of hours later I received the shocker 'You f***ing c***, you f***ing got me interested then just f***ing dumped me, I want to kill you, you c***' (I've given her the

> benefit of the doubt when it comes to grammar). Now, that scared the s*** out of me. Thankfully, she didn't know where I lived! Panicking like mad, I didn't know what to do. Thankfully she came to her senses and sent me a really re-assuring text 'Sorry about that, I have some anger management issues, but they're getting sorted out, sorry again'. Phew, I thought, that's the last of that! No! Then a few minutes later, a repeat of the 'are you sure text, followed by another 'f***ing c***' text, followed by another apology. After five days I counted over 500 texts from Arlene. Then all of a sudden it stopped. That scared me off internet dating for a while.
>
> *(Peter, UK, individual blog, 2005)*

In the early stages of the narrative we get a view of internet dating as 'pleasant' or 'nice'; it conforms to fairly heteronormative scripts. There is nothing very troubling or heterodoxical in these descriptions. Usually dinner is involved, romance is in the air. The narrator then appears to try to avoid being seen as naive by explaining that he did have some doubts about Arlene early on. Peter explains that it all 'started to unravel' when she 'started to put things into two camps. Fiction and non-fiction'. This data also indicates some unease amongst these men, about women controlling the pace at which a relationship proceeds. Arlene is described above as being 'very good at kissing'. Then we get the tale of her 'I love you' text. Peter's drink spitting reaction is presumably meant to indicate his shock and readers seem to be expected to appreciate how unexpected this was so early in their dating relationship. Previous relationships are also used as evidence to build a portrait of a 'nutter'. Peter mentions that Arlene is recently single 'after her boyfriend who used to beat her up and anyone who looked at her too, finally went too far'. It is also noted later that she was left 'with a mountain of debt' *(Peter, UK, individual blog, 2005)*. Peter also establishes himself as tolerant and willing to trust in noting that it was his relative inexperience with internet dating, the fact that 'it was still early days' for him, which explains why he 'naively' suggested meeting again despite his misgivings. What follows makes those misgivings appear justified, as the 'I love you' is presented as surprising for that early in their relationship and Peter explains a desire to go more slowly than Arlene. Only 'a couple of hours later' the narrator says he received a text from Arlene asking if he had made 'the right decision' followed another couple of hours later by 'the shocker' angry text saying she wanted to kill him. He said that 'scared the s***' out of him and, for a time, he was put off internet dating.

This captures the broad structure of the nutter narrative. *At first all seems well*, they have a 'nice evening', even a second date. There is some getting to know each other. However, the narrators usually try to indicate that they *started to 'get cold feet'* fairly early, due to some sign of a lack of truthfulness on their date's part and/or their date being sexually or emotionally forward, some perceived kinkiness or naughtiness, or because of some mention of problems in their past relationship/s. With Arlene, it is on their second meeting that the narrator says he became aware that the details Arlene has provided online included a considerable degree of 'fiction' and then it all began to unravel when she texted saying 'I love you' after three dates. The narrator also mentions the recent relationship with a violent boyfriend and 'a mountain of debt'. However, Peter gives her *the benefit of the doubt* and carries on, in this case 'naively' suggesting meeting again. He then explains how he began to feel uneasy (again) because 'things were *moving way too fast*'. Finally, the expletive filled texts followed by bombardments of angry and apologetic texts in turn are presented as *the revelation* of Arlene as a 'nutter', as mentally unstable.

At first all seems well: Pleasantness

In the initial stage of accounts the pleasantness described seems to be part of the cultural reinforcement of compulsory heterosexuality (Rich 1980). The 'nutter' narrative usually begins with some description of everything being fine when they were getting to know the person online and then meeting, although in the version above this is a very truncated account of the first two dates. A story of gradual getting to know each other is told in greater detail by Peter when he tells the tale of his first internet dating experience with another woman, Natasha, whom he meets on *Match.com*.

> We corresponded a lot, maybe even as much as 50 'letters' backwards and forwards. We had taken so long to do this that I felt like I knew every detail about her life and she about mine. It was time to move in for the kill and get that first internet date! I suggested we speak on the phone, we were both in France, after all. She tried to call me a few times and couldn't get through. I put it down to the telephone exchange in the local area, which was in a very remote part of the French countryside. I suggested that I call her. Then she went very quite [sic] all of a sudden, no emails, nothing.
>
> *(Peter, UK, individual blog, 2005)*

We get more of an indication of a developing relationship, although even here there are efforts to signal that perhaps something is not right by him implying that there was something suspicious about her not being able to get through on the telephone and the fact that she stopped contact 'all of a sudden'. A similar account is evident in Steven's tale of a date he went on in 2011, which began with them having a 'surprisingly fantastic first date' and seeming 'to click straightaway' *(Steven, UK, individual blog, 2011)*. Soon, he and other narrators tell us all is not so well.

Getting cold feet: Gender and control

When nutter narratives describe how they started to get 'cold feet' it reveals limits to the challenging of gender norms around men having control in relationships. 'Cold feet' are often traced back to initial doubts about truthfulness, but they are heightened by men's discomfort and women's fear when they think partners have been too sexually forward or when what they say about past relationships appears to depart from the model of strong men directing passive women.

In these narratives daters illustrate how the truth of what someone had shared online came into question. This is often signalled by comments about the photos they have seen online as too 'flattering' and not a 'real' version of the person. Complaints about 'minor deceptions' are common amongst internet daters, even if many users practise them (Heino et al. 2010: 435; see also Toma et al. 2008 and Chapter 3). Such complaints are evident again in the story about Natasha: 'I should mention that her photos were somewhat more flattering than the reality of what stood before me' *(Peter, UK, individual blog, 2005)*.

Apart from indications of deception like these, there might be some hints of odd or unpredictable behaviour. Steven's 'fantastic' first date with a woman we will call Nancy turns quickly odd, he tells us, 'when whilst standing outside a pub trying to get our bearings, she inexplicably just walked off without me, saying she had to get home'. Having missed his last train, he catches her up and she lets him sleep on the sofa. He decides to meet her again *(Steven, UK, individual blog, 2011)*.

The developing unease described by the phrase 'getting cold feet' is often linked with some indication that the male narrator thought their date was being too sexually and/or emotionally forward. It has been suggested that online dating preserves the offline rule that men initiate contact (Barraket and Henry-Waring 2008: 15, see also Chapter 3). Similar implied criticism appears to be behind the decision to tell readers the following about what happened when Natasha farewelled Peter at the airport: 'She kissed me quite intensely for a while, which was more than I was expecting and it was as if to say 'maybe there is something there' *(Peter, UK, individual blog, 2005)*. The 'more than I was expecting' suggests this kissing 'intensely' was rather sexually forward.

Sometimes, a predilection for sexual kinkiness of some kind starts to build a picture of the person as mentally unstable. This is the case with Steven, who fairly quickly gets into a relationship with Nancy, but notes that although they were 'getting on really well [he] had become very conscious of her severe mood swings and hyper-sensitivity'. The 'first sign of it' comes when he goes to her flat after receiving some 'dirty texts'. He explains that, 'as it turned out, during sex, she liked her hair being pulled and she also liked to be throttled and strangled I tried to do what she wanted but I just felt extremely uncomfortable' *(Steven, UK, individual blog, 2011)*.

The fact that the relation of this information about her sexual wishes is framed by his noting awareness of 'mood swings and hypersensitivity', seems remarkably close to nineteenth-century portraits of hysterical women as over-emotional and oversexed (Foucault 1990; Gilbert and Gubar 1980). There also seems some double standard on his part and adherence to fairly heteronormative views of sex when he is amazed that she throws him out of her flat after asking if he has any kinks and him replying he 'couldn't really think of any' except he watches porn sometimes. On his first internet date with Danielle, whom he met via *OkCupid*, another blogger seems to intimate that forward kinkiness on the part of women somehow transgresses heteronormative limits. He notes that during dessert at their pleasant dinner 'she reaches over, takes my hand . . . Takes the chopsticks out of my hand, starts sucking my fingers. WEIRD. But I didn't stop her. I should have' *(Jacob, UK, individual blog, 2013)*. The suggestion that he should have stopped her, and the capitalisation of 'weird' seem to indicate his discomfort.

Indeed, men's nutter narratives very often indicate that the women they are concerned about are somehow 'weird'. However, weird has a gendered tenor. Peter, for example, implies that Arlene's relationship history reflects something negative about her. He offers a similar reaction to Natasha's account of past relationships.

> I asked about ex-boyfriends and she went to some lengths to tell me about how mean Russian men were and how badly they treated their women and yes, she'd had a few boyfriends, but nothing for a while and nothing long term (this worried me, if a woman hasn't had a long term relationship by her mid-thirties, is she ever going to have one?)
>
> *(Peter, UK, individual blog, 2005)*

That Natasha had not had a long-term relationship 'worried' Peter and made him question her ability to do so in the future. His disquiet about her previous relationships is also revealed by the way in which he discusses how this 'nutter' may have related to men. Natasha tells him Russian men are mean and treat 'their' women badly. Readers are simply left to interpret what this might mean. In a similar vein, the London-based music fan tells us that Nancy 'let slip that she wasn't used to men saying "no" to her and that she always got her own way'. He goes on to say that she was explaining that him 'not bowing down to her every whim was a new experience for her' *(Steven, UK, individual blog, 2011)*. The implication in the case of both Peter and Steven is that men start to have doubts if they think that women have resisted 'bad' treatment in previous relationships or are too used to 'getting their own way' and exerting control in relationships.

By comparison, the women's accounts are more likely to mention concern about men being aggressive and arrogant. Men's early sexual forwardness is often seen as an indication of these problematic possibilities. Such forwardness may even happen before meeting, as one woman explains: 'I recently received a message from a manboy on [dating site] that firstly listed his qualities, among which he included good kisser, good lover (yup odd already and he did not include arrogance amongst his virtues)' *(Charlotte, UK, general blog, 2014)*.

Similarly, Lucy often notes in her blog that she does not like too much messaging or texting about sex early on and she records that one date's 'overbearing keenness did make her feel a more than little uncomfortable' *(Lucy, UK, individual blog 2019)*. Another woman blogger also notes receiving an email from a man whose 'profile said he was creative and calm, with a kind and gentle soul. All incredibly good qualities to have, [she] thought, even if it's by his own admission'. She describes seeing 'lots of potential' in this suitor, but then notes that his next emails took her 'quite by surprise' when he interrogated her about saying on her profile that she does not want children *(Sonya, Australia, general blog, 2013)*.

The benefit of the doubt

Narrators respond to the doubts that cause the 'cold feet', by emphasising that they are nice, tolerant, trusting, reasonable people, but how they do this is gendered. The men's presentation of themselves as tolerant seems to emphasise the reasonableness of males in contrast to their dates. These women are presented as untrustworthy and potentially mad. Given this presentation, the men narrators not surprisingly appear to feel they have to explain why they met and carried on seeing such women, despite the doubts they had. In his first internet dating experience Peter implies that he wanted to trust Natasha, despite concerns emerging early on.

> We talked about her time in Paris in London and I got the distinct impression, from the fact she couldn't remember where she had lived in Paris that she may not have been telling the complete truth about that part of her story. Anyway, I gave her the benefit of the doubt.
>
> *(Peter, UK, individual blog, 2005)*

The doubt about whether her story is 'true' is put aside, there is a desire to not be overly suspicious and to assume that the potential partner is to be trusted. This is also clear as the story unfolds. After ceasing communication 'all of a sudden' as described above, Natasha gets back in touch.

> A week or so later I received an email from her saying that she had had to return to Moscow at short notice and that she would like to talk again. So, taking this at face value, I arranged to call her in Moscow. The line was very bad and we kept getting cut off. We finally had enough conversation pieced together and we agreed to meet. We started talking about where and

when. She offered London and I said that sounded good to me, but then she had problems with her visa.

(Peter, UK, individual blog, 2005)

Women's untrustworthiness is one sign that they are not rational and not in 'proper' control of their emotions. This is an old trope, recognisable in much literature in the figure of the madwoman in the attic (Gilbert and Gubar 1980) and in science in the guise of the hysterical woman (Foucault 1990). In the above quote Peter signals suspicious signs. However, despite these signs, Peter (somewhat graciously, we are presumably meant to think) decides to trust Natasha. Why the sudden return to Moscow? Well, he took it 'at face value', he believed her.

Likewise, Steven *(UK, individual blog, 2011)* notes that after Nancy threw him out for admitting he watched porn, he 'managed to patch things up with her, although in retrospect [he] wish[ed he] hadn't'. After other similar incidents, where he describes her being angry and often verbally abusive, he says he 'always like to give people another chance and to sort things out in an adult way' and that 'she had so many good qualities' which was why he 'kept wanting to give her the benefit of the doubt'. Peter and Steven continuously remind us of their measured and generous responses as they return to contacting and dating women they increasingly do not trust.

By contast, in the relatively few women's accounts of 'nutters' encountered in online dating, the women seem to mostly weed them out at the messaging/emailing/chatting phase. The women are comparatively less likely it seems to ignore suspicious signs. Nevertheless, a man's sexual aggressiveness can sometimes take time to emerge—even in this early phase. When problems like sexual agressiveness arise, the stories women narrators tell begin to share some of the characteristics of those men. For example, women narrators also feel the need to provide an explanation of why they continued these 'conversations' as long as they did. And, sometimes they do also meet up with these doubtful men. In Lucy's case one of the stories she tells about a 'strange' man she meets while in Africa suggests that normally she 'would baulk' when in online chat he says he 'wants to "bribe" her with dinner' rather than meet for a coffee on the first date: 'there's something quite attractive about Adil's flirty confidence. And This Is Africa, after all, where apparently she's more of a risk-taker, more confident, and where everything seems to be larger and faster and more dramatic. So she accepts' *(Lucy, UK, individual blog, 2018)*.

Lucy (using the third person to refer to herself) attributes her overcoming of her doubts to being in Africa, where she feels the context is different and she is different. After meeting up with this 'strange' man, who is Asian, she also wonders if it is 'cultural differences that mean they don't gel quite as easily?' Nevertheless, despite this meeting making her think 'there was definitely something slightly strange about him', she tries to explain it away as because 'he was just nervous' and 'he knows she's only here a short time, so no harm in keeping him around to go out for nice drinks and dinner every now and again'. Her racialisation of this Asian man is brushed over by presenting herself as cosmopolitan, relaxed and reasonable, but this has a slightly different effect than the 'tolerance' the men display. Where the male narrators' emphasis is upon them altruistically giving women 'the benefit of the doubt', Lucy offers an account of trying not to overthink and being laidback, convivial and cheery. She does not want to appear to be a nutter.

Overall, the exemplary women's narratives we have chosen can be read as attempts to avoid being seen as demanding, complicated or unstable and thus having to be measured and good-natured, even where the men are forward. The nutter narratives largely diverge in relation to gender from the point of 'cold feet' doubts arising and continue to show signs of this divergence when the narrators explain their responses to suspicious signals. The next section shows what these narratives can tell us about some of the complexities of negotiating the speed at which internet dating relationships proceed.

Moving way too fast

The male 'nutter' narrators begin to register the alleged craziness of their date through presenting the women as overly 'forward' in seeking emotional intimacy too rapidly. The farewell scene with Natasha at the airport continues with Peter telling readers that 'just before I went to the gate she said, if we are going to see each other again, it will need to be after you divorce'. This is meant to indicate a too rapid seriousness on her part which he did not share, as he reveals barely a line later that, 'I didn't really see a future for Natasha and I'. In Steven's story of his relationship with Nancy, he deems her guilty of this over-seriousness even after several weeks of what he himself describes as a fairly intimate relationship with them in 'constant contact'. After one or two accounts of her sensitivity and 'inexplicably' moody huffs, Steven *(UK, individual blog, 2011)* notes that she goes overseas and for 'the three weeks she emailed every day telling me how much she missed me. He found this 'bizarre and uncomfortable (which I told her) as we hadn't been going out long at all'. Again this seems to indicate that she is getting ahead of herself.

For these men 'moving too fast' is alarming and this indicates limited departure from the gendered heteronorm of men being the initiaters of contact and the ones who decide when a relationship can proceed emotionally to the 'next level'. The assumption that men will be the initiators does not seem to have been especially disturbed by a shift online. All the same, while there is significant research to indicate that men initiate contact at far higher rates than women, it has also been suggested that around 20 per cent of first communications between online daters come from women (Dinh et al. 2018; Fiore et al. 2010; Su and Hu 2019). It is possible that proactive women are more likely to use online dating, or it may be that internet dating at least allows women greater control over the process of meeting men (McWilliams and Barrett 2014: 414). We examine the latter possibility.

The women whose stories we have collected do mention finding internet dating a useful way to deal with the 'forwardness' of men, which can be more difficult face to face. Often, in our sample of nutter narratives, the forwardness is directly sexual. One of the men bloggers relates the story of a woman friend where the internet date does not go well:

> she heads home, thinking, what a douche. He texts her, saying 'I realise the date went badly, but was wondering if you were still interested in sex. I have a massive penis. Bob. XX'. And attached, is a picture of said massive penis.
>
> *(Jacob, UK, individual blog, 2013)*

This might seem an unusual or even fairly extreme example of a response to an unsuccessful date. However, it is hardly uncommon (Anderson et al. 2020; Paasonen et al. 2019; Thompson 2016; Waling and Pym 2019). Even when heterosexual men are not quite this forcefully blunt about their intentions, women daters experience considerable discomfort when men seem to leap too soon to declarations of romantic, emotional interest. Lucy is wary when Adil disappears before their second date and, again when she asks why, he messages that 'nobody could ever like you or take care of you the way I could'. As he sends what she thinks are increasingly strange and intense messages, Lucy *(UK, individual blog, 2018)* speculates about whether 'he's just entertaining himself by trying to mess with her head' and posts screen shots of her responses, reminding him they have 'only had one date' and being reasonable and explaining to him what she found odd. She ends by appealing to her readers: 'I know. I have no idea what just happened either'. They will surely recognise her sanity and his weirdness. And readers will understand that a woman needs to be cautious. In offering to take care of Lucy her date is invoking norms around men as protectors, but doing this so early after one actual meeting appears intrusive and domineering rather than indicative of emotional depth. Thus, Lucy seeks help from others in navigating the shifting boundaries around what is normal and acceptable behaviour for men.

Managing the pace and manner of their engagement in face-to-face encounters is necessary given women's realistic fears of stalking and violence. However, online 'meeting' can help women to filter out men who move too fast. For women, inappropriate 'intimacy' relating to their bodies may appear in the emailing stage, but there it may be easier to address. The suitor whose emails took Sonya, an Australian woman blogger by surprise, moved straight from 'initial pleasantries' to asking why she had 'listed "undecided" next to [her] desire to have children' *(Sonya, Australia, general blog, 2013)*. In this narrative this is in fact the revelation, followed immediately by an explanation giving him the benefit of the doubt by saying 'that he "immediately apologised for being blunt, but justified this approach by saying he believed in being direct"'. Nevertheless, she then describes how 'her first reaction was to place both hands protectively over [her] womb'. She indicates that she thought this was too much, too soon by asking readers if 'that [is] really an appropriate question for someone to ask you before they've even met you?'. She highlights this further by saying that she thinks discussing whether she wants children or not is 'too much to consider when I don't even know the guy's surname'. However, she suggests that she learned from this that being clear about what you want will mean 'you won't waste valuable time'. Thus, a contradiction emerges between not 'wasting time' and not 'moving too fast'.

The dilemma around when to become more intimate, while a feature of discussions by both men and women online daters, produces gender differiented reactions. What is interesting here is that while female internet daters/bloggers will outline stories of avoiding 'nutters' which illustrate women attempting to assess men's behaviour in a discriminating fashion, there are a range of associated forms of advice in which they are also castigated for being overly particular. Women are warned that they may miss the boat! In this context, in her Australian online advice column Bettina Arndt approvingly quotes the American best-selling therapist, Lori Gottlieb, who famously promotes settling for Mr Good Enough and declares that 'women themselves are at risk of ego-tripping themselves out of romantic connection'. Gottlieb points out that she 'made a mistake not looking for a spouse in her 20s, when she was at her most desirable . . . [but now] advises thirty-something women to look for Mr Good Enough before they have even less choice'. Once they are 40, Gottlieb intones, good men will choose those desirable younger women and they will be left with the dross (Arndt 2012; see also Groves and Chan 2017).

The tensions between going too fast and going too slow are there for both men and women daters, but only the latter are regularly reminded of the awful consequences of potentially missing out. In that case, how nutty a man should be accepted? What to do? Some decisive revelation is often cited by internet daters as helping make decisions to not continue further. In most stories, whether by male or female narrators, after the 'moving way too fast' part of the story, the revelation of nuttiness is not far away.

The revelation: Beyond the normative

When it comes to the revelation of nuttiness, we see where daters have been judged to have gone beyond acceptable normative boundaries around gender. They have stepped out of the acceptable limits of femininity or masculinity, and are discarded as potential partners. Narrators appeal to the audience to reinforce their revelation as obviously showing that person as beyond the pale in terms of what 'we' all would find acceptable. Thus gender norms are policed, but again in ways that vary between the men and women narrators.

The men reveal the women as frightening and freakish, or sometimes bring them back into gender bounds by portraying them as looking for (rich) husbands. Peter's story about Natasha is missing a revelation and in that sense is not a classic 'nutter' narrative. The story fizzles out with Peter explaining that they 'kept in touch for a while and it became clear to me that divorce was a big thing for her, separated was not enough'. There is a brief account of her calling a couple of years later asking for some business contacts, but that is all. Peter was not rushing to get divorced for a woman he has portrayed as dishonest. She doesn't seem to be wife material by his account. Yet, revelations do not need to be dramatic to position the women as bad or mad. With Nancy, Steven (an internet dating veteran of about seven years at that point) starts the story by saying she is 'the most mental and scariest person that [he has] met so far on [his] internet dating travails'. The revelation is more a cumulative account of her 'moods' and a story of a final falling out when she returns from her overseas trip. He finishes the story by saying,

> I can't emphasise enough how much in the end she actually scared me with her mood swings, and I'm not sure if I was dreaming it or not but I have quite a clear memory of seeing her going through my mobile phone on the last morning I was with her.
>
> *(Steven, UK, individual blog, 2011)*

Sometimes the nutter narrative may skip some of these stages and a revelation may more quickly emerge to draw a line between 'normal' and not to be tolerated ways of behaving. With Jacob, the blogger who met Danielle for dinner, the story quickly arrives at the dessert, when she reached over and began sucking his fingers. At this point, the revelation appears:

> [s]he then bit my fingers, really hard! I screamed at the top of my voice (default setting: loud) 'WHAT THE FUCK DO YOU THINK YOU ARE DOING?!', causing a scene in the restaurant. She tells me she thought I'd like it—then bursts into tears. I wrap my bleeding hand in a makeshift bandage I fashioned from a napkin, walk to the bar & say 'I'd like to pay & leave, please'.
>
> *(Jacob, UK, individual blog, 2013)*

The story continues to tell of how the bite wound became infected and as this is 'one of the worst dates ever' it prompts him to start writing his blog because as 'one of the few (relatively) normal survivors out there' he felt he could provide a 'survival guide' *(Jacob, UK, individual blog, 2013)*. Thus, as Goffman's approach to interactional scripts suggests (1968), the narrator's own status as a 'normal' is reinforced by portraying his date as 'weird'. Indeed, albeit in joking fashion, Jacob describes internet dating generally as 'a wasteland populated by freaks, huge cocked mutants and blood crazed biting cannibals' *(Jacob,*

UK, individual blog, 2013). Such out-of-place sexual forwardness can place a woman in the territory of mad and bad.

Women's accounts may also proceed quickly to the revelation, but the boundary is between charming masculine directness and unappealing macho misogyny. When discussing a message from a 'manboy' one woman blogger reveals that,

> the most worrying part . . . came later when he switched to caps in the fourth paragraph to proclaim 'and above all she must be OPEN AND HONEST, I DO NOT PLAY GAMES AND I EXPECT THE SAME FROM HER!!!' Okay psycho . . . He rounded off by saying he was looking for a 'down to earth normal relationship' . . . sure. . . .
>
> *(Charlotte, UK, general blog, 2014)*

Heterosexual men may not necessarily be extreme patriarchs because they want to avoid women with unpredictable behaviour, but where they draw the line between exciting and too forward is revealing in terms of how shifts towards the heterodoxical, towards unsettling heterosexual gender norms, might be 'straightened up' or stopped. These women also might reinforce heteromasculine forms of behaviour, for example by expecting men to make the first move, an expectation which is seemingly usually fulfilled (Bruch and Newman 2018; Kreager et al. 2015). However, as West and Zimmerman (1987) point out in their landmark analysis of 'doing gender' (1987), gender display is an essential factor undergirding how men and women interact. Women's caution regarding relationship progression may not only indicate their reiteration of heteromasculine norms, and the greater selectivity of their responses to male attention (Bruch and Newman 2018; Sassler and Miller 2011),[3] but also that they may not have much space to do otherwise without being thought mentally unstable and/or sexually available/excessive in terms of gender norms.[4]

Women's nutter narratives suggest that they value the opportunities internet dating provides to filter out at an early stage those men who are sexually aggressive or potentially misogynistic and/or violent. Boundary lines are thus drawn. In particular, it seems that women can draw boundaries online, where they appear to feel better able to control encounters. In this respect, internet dating may, initially at least, provide women with greater power over dating progression than offline meetings.

What does this tell us about gendered heteronormativity in internet dating?

Nutter narratives help us see the boundaries of gender innovations in heterosexual internet dating. They show how departures from gendered heteronorms might be policed and how and why internet daters might shrink from doing anything which seems to go beyond subversion. For men, they may fear loss of power and control if women take the lead. For women, they may fear being considered mad (or madly sexual).

The narratives people tell about their internet dating experiences can reveal how shifting yet stubborn heterosexual gender relations shape those experiences. We have argued that the nutter narrative is a commonly told story that exposes many of the gendered assumptions and ways of interacting that can reinforce inequalities between women and men. It is a narrative that helps us understand where the limits to gender heterodoxy sit and how they are guarded. The nutter narratives suggest that the gender innovations enabled by internet dating may travel out from the heteronormative centre, but not too far. What we offer in the rest of the book is an analysis of what kind of innovations are possible, but here we get to grips with the outer fences, the lines in the sand, beyond which it is dangerous to go. Technology has affordances, but the internet is not outside of regulatory power. The nutter narrative is one mechanism via which that power is exercised and gendered selves and interactions produced.

In the beginning nutter narratives indicate that heterosexual relationships in their early stages may feel pleasant, as long as they do not trouble gendered heteronorms. As these narratives proceed we see how women still have to navigate tacit norms that men should control the initiation, pace and progress of relationships. This is informed by sexual double standards in which men's 'cold feet' seem to echo concerns about being tricked into, or 'trapped' in, relationships. It may thus not be surprising that these men often police any efforts women may make to assume control, but what seems less well acknowledged is that what and how we trust and doubt may be highly gendered. This would benefit from further research. While both men and women suggest that they like to give others the benefit of the doubt, the form and limits of this tolerance are gendered. The men's narratives draw on long-standing tropes of mad, bad women against whom they appear as tolerant, generous and reasonable. By contrast, the women's stories about nutters can sometimes sound like those of virtuous maidens protecting themselves from aggressive, predatory males, although they may also hint at women's attempts to maintain independent agency.

The narratives suggest these men fear and stop women from moving too fast towards commitment and that women try to slow men's rush to physically meet and potentially to have sex. There are glimpses in this data of differences in the point at which men and women might veto other further intimacy, with some examples of women exercising that veto in the online stage of communication and men later. Such examples suggest strongly that women are able to make use of the online early stages of dating as a kind of shield, a shield which becomes more difficult to weild when in the face-to-face presence of their male dates. Gendered power relations in heterosexuality may be more negotiable at a distance, and in this limited sense online dating offers some innovative possibilities, even if they do not escape the bounds of heteronormative prescriptions for protective men and pure women.

There is also something interesting in how these nutter narratives come to a revelatory epiphany. Weirdness can reinforce the narrator's presentation of himself as a normal man trying to avoid crazy (over-sexed?) women. The comparatively fewer women in our sources telling nutter stories, meanwhile, seem concerned to keep some control over their bodies and when and how they respond to men's advances. They might seem to be delaying sharing their favours in contrast to women labelled 'nutters' who seem to be rushing to bestow them. This however reiterates madonna/whore binaries that we would be eager to see dismantled. Whether women narrators are 'good' and whether the women so labelled are 'really' nutters—and whether this makes them 'bad'—is not the point. As characters in this story, women 'nutters' seem to serve a purpose for men in navigating between explicit misogyny and implicit reluctance to accept women as equal partners in driving relationships forward. Meanwhile, women have to read and respond to online communications carefully in order to avoid men who might be dangerously heteromasculine and find, if that is what they seek, those men who are within normative territories that promote equal partnerships. The revelations of craziness show that those women and men who are deemed to have strayed too far from gender orthodoxy are likely to be shunned and shamed. Let these monsters be a lesson to others, the narratives seem to say.

In sum, for the most part the nutter narratives we encountered were most often employed to cast some heterodoxical deviations from gender norms as implying inappropriate, untrustworthy and potentially risky dates. The narrative especially targets shifts in highly gendered norms around 'forward' behaviour (particularly sexual behaviour). The nutter narrative works to police the recuperation of traditional expectations of gender and heterosexuality, thus framing any innovations as primarily negative. All the same, it should be noted that the nutter narrative mode of policing frequently exists alongside more positive or at least open-ended accounts of these novel variations, perhaps confirming the emergence of growing uncertainty about social norms of intimacy rather than any straightforward resurgence of gendered heteronormativity. In this context, it is worth remembering that the nutter narrative in internet dating does not typically result in users giving up dating online and is frequently employed as a means to clarify what users really want from dating. While the narrative may well often provide instructional stories directed towards risk avoidance and the reiteration of differential gendered expectations, it also shows that boundaries and limitations around gender are capable of change and are continuing to alter.

Gender orthodoxy online is by no means set in concrete. Many women take up mechanisms available to them through the online technology of intimacy which enable them to filter their dates and thus avoid aggressive/sexual pressure. While in many ways this is an extension of heterosexual women's longstanding role in acting as sexual gatekeepers, it does also enable them to assert some heterodoxical divergence from norms of feminine passivity in relationship initiation, pacing and progression. Some women do initiate contact to meet male dates, despite the possible risks associated with transgression of gender norms that they may incur. Although there are constraints on innovations, new norms do emerge as the following chapter demonstrates.

Notes

1 We note that the term 'queer' implies 'an opposition to hegemonic norms, a protest against the "normal"' (see Showden 2012: 8 and Chapter 5 for further commentary).

2 Nutter narratives are closely aligned, though usually raising somewhat more concerning behaviours, with the seemingly endless material on 'red flags'—that is, advice regarding signals about what daters should avoid (see for example, Dodgson 2018).

3 Bruch and Newman's American study (2018) found that,

> a defining feature of heterosexual online dating is that, in the vast majority of cases, it is men who establish the first contact—more than 80% of first messages are from men in our data set. As a result, there is little information about women's aspirations contained in first messages. On the other hand, women reply very selectively to the messages they receive from men—their average reply rate is less than 20%.

An eHarmony study in Australia reiterates this pattern (Croffey 2016). According to OkCupid data analysis, this marked tendency for women to not initiate contact holds across age and no matter how many messages a woman receives (Cooper/OkCupid 2016).

4 Women appear reluctant to initiate first contact in internet dating, despite considerable evidence (and online dating advice encouragement) that they are likely to receive many interested responses if they do so. Perhaps women judge that they may attract interest but not interest that they want.

Chapter 3

The new norms and etiquette of internet dating

Internet dating itself has become normalised in many countries, and heterosexual internet dating is to a considerable extent part of an apparatus for replicating intimate norms concerning 'coupledom' (Roseneil 2005; Wilkinson 2012). Nevertheless, there remain some uncertainties about the norms of internet dating, as is so with using all new technologies (Baym 2015: 134). Those engaging with it have to ponder general questions about when it is appropriate to use it, who to communicate with and how much information to give as they proceed (Baym 2015: 134–136; Chambers 2013). This entails more detailed decisions about how to present themselves appropriately, who should make the first online contact, how and when to propose a telephone call, when it is time to meet in person and what to do when first meeting offline (Harper 2018). There are questions about whether it is polite to reply to all the messages received, or how to filter them if there are many. Is it acceptable to be in contact with more than one person at once and when and how should you tell people 'thanks but no thanks'? Common sense and face-to-face rules of interaction might help, but they too are often uncertain and the technical affordances and sheer volume of interactions of internet dating seem to demand new norms and rules about how to act (Hardey 2008; Heino et al. 2010). What are those norms and how are they produced and navigated? We focus in this chapter on what are broadly accepted practices and attitudes and ponder whether these emerging 'customary' understandings recuperate hegemonic gender and sexuality stereotypes or offer any possibilities for more egalitarian innovations and, if so, to what degree?

Internet dating is often characterised as a novel realm involving new rules and practices, but academic attention to its norms and etiquette has been limited. Despite two decades of investigation of 'netiquette', much of it has been descriptive (for example, Scheuermann and Taylor 1997) and links to wider social changes in relating have only rarely been made (Beasley et al. 2018; Beasley and Holmes 2016; Chambers 2013: 121–141; Helsper and Whitty 2010; Holmes 2011). Rather than examine changes accompanying digital dating as examples of Giddens's account of the emergence of 'plastic sexuality' (Chambers 2013: 121–122), we explore the development of new online dating norms as less individualised instances of innovations in social relations. The ways in which people in interaction navigate the technical and actual possibilities of internet dating are characterised as socio-historically situated, embodied and emotional. Although people may not always follow norms, the evolution of etiquette can help reveal shifts in how things are done, as the emergence of new norms often involves describing and denigrating past practices (Elias 2000). We propose that users also engage in participatory negotiation and navigation of new norms. They do this via ethical considerations and discussions around how to do internet dating which are collective as well as individual.

In this chapter, we set out how netiquette can be useful for identifying social norms and examining changes in them that are divergent, transgressive and subversive. We then argue that much ethical norm making is no longer fed down from elite authorities but can be done in a more collective and participatory fashion via the internet. We consider what some of the new norms might be by examining to what degree they may depart from previous social rules. We present findings that suggest norms around honesty are in actuality diverged from in making profiles and that there is some transgression of norms about isolated searching for 'the one'. Furthermore, we extend examination of the supposition in Chapter 2 that a degree of subversion occurs around previous norms of men pursuing women. We then discuss to what degree rather more heterodox pathways may be discerned. Transgression or subversion are evident in breaking from heterosexual gendered norms about who pays for dates and around first having sex. We consider why it is important to set out these norms and we make sense of what they tell us about norm making as a relational activity and one which can be aligned with or even produce social change.

Etiquette as a marker of changing social norms

A study of pronouncements on etiquette can illustrate how social norms and practices alter in relation to large socio-historical processes about which we have 'very little direct information' (Elias 2000: 72). Norbert Elias (2000) famously established this in his book *The Civilizing Process*, which was based on analysis of manners books setting out guidelines on 'proper' behaviour (Elias 2000: 71). This analysis underpins his argument that, in the European context, norms became subject to increasingly formal regulation from the eleventh century up until the nineteenth. As societies develop, argues Elias, there are more norms and stricter rules about how to behave. More formal manners are expected as modernity emerges, assisted by the introduction of new technologies including everything from forks, to handkerchiefs to toilets. Rationalisation processes subject individuals to more regulation as more unrestrained forms of behaviour gradually become seen as unreasonable. Individuals, including children, are increasingly expected and socialised to exercise increasing self-restraint over their bodily and emotional needs. Relatedly, there is a change of consciousness which involves placing greater emphasis on individuals and an associated decline in more communal practices such as people drinking from the same cup, and eating from a common dish. This account of regulatory individualisation differs from the approach of writers like Giddens, Beck and Bauman whose emphasis on individualisation is concerned with shifts away from kinship and local ties and towards the autonomy of the self (Holmes 2011). Elias (2000) also asserts that the state monopolises the 'acceptable' use of violence as part of the process of pacification.These sub-processes of rationalisation, individualisation, socialisation and pacification constitute for Elias 'the civilizing process'.

While aspects of an Eliasian approach to the development of norms may be useful with regard to the novel practice of internet dating, it also has limitations. The emphasis on psychogenetic as well as sociogenetic elements of social change has advantages for examining feelings, but Elias's reliance on Freud produces an understanding of 'natural' human drives as 'repressed' by the civilising tendencies of increasing societal complexity. While Elias can be read as arguing that society modulates rather than represses drives (Burkitt 2012: 41), this 'modulation' is presented largely as taking the form of prohibitions. In regard to sexual functions, for instance, Elias (2000: 160) argues that,

> [m]ore and more people keep the functions themselves, and all reminders of them, concealed from one another. [And] . . . the psychic structure of people is also transformed. The prohibitions supported by social sanctions are reproduced in individuals as self-controls. The pressure to restrain impulses and the sociogenetic shame surrounding them—these are turned so completely into habits that we cannot resist them even when alone, in the intimate sphere.

The resulting assumption is that the social world stops people feeling as they 'naturally' would. Drawing on Foucault's (1990) criticism of the 'repressive hypothesis', it is possible to consider that social life might subject people to feeling and acting in particular ways. Some ways of behaving may attain a certain discursive dominance but considerable ambivalence exists, perhaps especially around emotionalities. However, we suggest that the changing emotional and other social practices involved do not necessarily or always progress towards more civilised, socialised, rationalised and pacified social relations. Nor is it self-evident that modern sociality off or online is necessarily and consistently moving towards individualisation (Beasley 2017b; Jamieson 1999; Smart and Shipman 2004). There is also need to focus not just on social discourse but on practices and structural context (Burkitt 1998). With internet dating, for instance, it is not inevitable that it heralds a new dawn of ordered, clear, and conflict-free dating in which old rules, collective presumptions and social hierarchies are left behind. Indeed the rise of internet technologies concerned with social interaction, including intimate relationships, has lead to considerable evidence of and discusson about the development of 'trolling' following enhanced capacities to undertake censure and vitriolic attack from a distance. There has also been discussion of the development of 'echo chambers' in which such technologies enable the relentless repetition of group views that are by no means necessarily free of established hierarchies or rules (Bradshaw and Howard, 2017; Johnson 2017: 2–5; Megarry, 2014; Quattrociocchi, 2016; Rudder, 2014b; Usher et al., 2018). Elias's account of the direction of social change as 'civilising' appears as overly linear and gender biased when considering the digital age. As Pasca (2018) points out, '[w]hen users log onto dating apps, they do so without shedding their pre-existing conceptions about how the world works. It's unsurprising that the interactions users have in these spaces come with the baggage of real-life misogyny'.

Internet dating rules and norms do however need to be understood within the context of social changes that have seen at least some alteration in gender, sexuality, class and ethnic relations. The increasingly formal regulation of behaviour Elias describes as occurring from the eleventh until the nineteenth centuries, has arguably been replaced by an informalisation from the twentieth century (Wouters 1995a, 1995b, 2008). Etiquette since that period can be interpreted as illustrating a process in which diverse 'regimes of manners' emerge that require more shifting and reflexive approaches to regulation (Wouters 2004: 210). New norms reflect how status barriers may in some ways become more uncertain or even blurred and people find they are required and even expected to become knowledgeable about emerging regimes of practice. Etiquette may function partly 'to draw and maintain social dividing lines' (Wouters 1995a: 108), but as those dividing lines shift and sometimes become less apparent, manners still retain significance because considerable anxiety may arise around how to behave. Thus, even if Wouters's (1995a, 1995b; 2008; 2004) thesis regarding increasing *self-regulation* in the contemporary world is arguable, growing informalisation and diversification of manners can mean more conscious attention to social parameters requiring flexibility and reflexivity in relation to both self and others, especially within novel social contexts. As we shall see, internet dating is one such context.

As 'the internet has become another mode of communication in people's everyday lives' (Helsper and Whitty 2010: 916), it has required new manners, or 'netiquette'. This has happened before as, for example, people learned to use telephones (Fischer 1994: 60–85, 183–185), but the speed of change surrounding netiquette is noteworthy as how people interact online has been altered with the advent of the more interactive and networked web 2.0 (see Beer and Burrows 2007). Some advice on email etiquette from the 1990s may still have relevance when it cautions users to consider the permanence of messages, the recipients, and to 'avoid responding while emotional' (Sproull and Kiesler 1991: 54). Other tips like 'do not insult or criticise third parties without giving them a chance to respond' make little sense within the frame of social media where those parties may be able to see and comment (Sproull and Kiesler 1991: 54) Like social media, internet dating is a more leisure-oriented activity than email, but this in itself illustrates significant alterations in people's relationships to computers. Academic analysis of those changes contrasts early concern about online relationships as encouraging 'uninhibited and aggressive' communication, with some more recent consideration of how the internet facilitates meaningful relationships, including romantic ones, despite a return to disquiet in current debates around phenomenona such as trolling (Baker and Whitty 2008; Baym 2015; Helsper and Whitty 2010: 916; Mantovani 1994; Miller et al. 2016; Sproull and Kiesler 1986; Valentine 2006; Whitty 2008a).

New norms develop as people navigate internet dating as a technology and a new social practice, but those norms and how they come to be has not been subject to substantial academic analysis. The little analysis that exists tends to argue that there are some novel edicts around how fast online communication should proceed and how soon it should become intimate, but that mostly face-to-face etiquette applies and old gendered patterns of interaction are reproduced (Barraket and Henry-Waring 2008: 159–160; Hardey 2002; Hardey 2008; Jamieson 2013). However, we want to examine these conclusions more closely as there are some recent signs of social innovation arising from the expansion of intimate connections online. For example, Hergovich and Ortega (2018) note that, in the USA at least, over time both heterosexual and same-sex relationships are more likely to have begun online. This has resulted in dramatically increasing the numbers of people who form couples without having any previous connections with each other and relatedly is associated with a sharp rise in interracial marriage. There may well be a growth in intimate relationships between those who were previously complete strangers (Rosenfeld and Thomas 2012) and perhaps some lessening of status differences (such as in relation to race) within social life generally, but such developments also bring uncertainty about what is acceptable, where and with whom (Wouters 2004).

This means that 'feeling rules' (Hochschild 1983) are not always clear and other taken-for-granted social discourses and norms may also wane as a guide to actions (Holmes 2010: 144–145). Thus, engaging in internet dating requires at least some improvisation, impression management (Frohlick and Migliardi 2011; Gibbs et al. 2006; Hardey 2008; Whitty 2008b), and reflexivity. Some internet daters may feel that there is a clear way to 'do the right thing', while others think there are a variety of conventions. Their own approaches may vary from the game-like to more serious wooing (Hardey 2008). It is likely the case that internet dating both reproduces existing norms as well as constructing new ones (Barraket and Henry-Waring 2008: 160; Hardey 2008; Jamieson 2013). Daters are not simply free to do as they wish online and there are collective and culturally specific, as well as individual considerations regarding how internet daters *should* behave (Island 2018).

Relational ethical norm making

There does seem to be some uncertainty about what the etiquette is in internet dating and in this context users are required to be reflexive, but they are reflexive in relations with others. One man noted that having not dated in a decade he 'didn't really know the correct form for approaching women' and 'got it wrong so many times' *(Peter, individual blog, UK, 2005)*. Gradually, through interacting and often getting it wrong he finds his way. The reflexivity exercised involves others and sometimes the internet daters in our sample act as ethical resources or guides for each other. They may, as one blogger mentions, warn about a particular dating site by describing it 'in very negative terms' *(Jacob, UK, individual blog, 2013)*. They may even give warnings about other daters, as when a previous date, 'kindly sent [Steven] a message' cautioning him to take care with a woman he was dating who she knew as having 'too many dramas' *(Steven, UK, individual blog, 2014)*.

New norms are crafted in interaction. There is considerable online reflection and collective sharing about what is and is not appropriate internet behaviour (Holmes 2011) including around internet dating (Hardey 2008), and in looking at this process we can see how spaces and opportunities for social innovation emerge. Some voices may be more dominant, but change, or the desire for it, is evident. One woman blogger, for instance, advises men in ways that seek to point out the unappealing nature of laddish and sexist forms of masculinity. She notes,

> in my virtual travels, as well as those of my friends, I have noticed some major mistakes that these love or most probably shag-seeking lads are making in their online etiquette*. Simple no nos that could be so easily rectified. So I have decided to outline a few here, all of these are made in the initial stages of contact or even prior to that, on a profile page. . . . Seriously boys, whatever you want from your foray in to the virtual flirtosphere, these fails will certainly render you less fuckable.
>
> *(Charlotte, UK, general blog, 2014)*

She goes on to advise them not to use selfies, to not display pictures of themseleves with champagne or sports cars that imply they are rich, to have photos other than ones where they are clearly drunk, to not send penis shots and to 'leave a little to the imagination' when writing their profiles. She also recommends they be 'careful with bringing [their] agro baggage/misogyny out too early'. It is also suggested that they use humour when sending the first message. This covers a range of advice about composing profiles to sending messages, and these online accounts often contain etiquette pointers on the whole process of internet dating. The following sections organise that advice in order of the common way dating progresses, to indicate what kinds of new norms might be emerging and being challenged.

Divergence: How honest should a profile be?

Divergent practices are those that are linked to the norm, but challenge its boundaries, as is the case with writing profiles where norms around honesty and the importance of looking good conflict. It is not straightforwardly evident that people are more likely to lie online, although this is sometimes claimed (Baym 2015: 106; Conger 2011). However, some tweaking of self-representation does appear common. Considerable advice is given to others about the importance of honesty in profiles but not everyone learns that 'you can't judge a book by its cover' *(Kali, woman, Australia, individual blog, 2016)*. Previous research suggesting that the internet may free relationship formation from the emphasis on (beautiful) bodies seems to have been challenged by work on internet dating that highlights how crucial photographs are in gaining the interest of other daters (Frohlick and Migliardi 2011; Whitty 2008b: 1714). In fact the importance of conforming to norms of attractiveness seems to make it common for profiles to contain minor deceptions about looks, relationships, age, weight, class, income and interests. The ethics of this is interesting because in relation to themselves people feel that it is acceptable and indeed necessary to exaggerate and deceive to attract others, but also—as Goffman notes—will judge other people as immoral if they know they have misrepresented themselves (Ellison et al. 2012; Toma et al. 2008; Whitty 2008c). Thus, it is common for online accounts of internet dating to rail against 'dishonest' presentations of self of various kinds (Hardey 2008: 1123). In the case which follows, a man's deception is discovered because the woman viewing his profile already knows him in real life.

> Last week a 'suitable match' sends me an email. 'Bob' from Paddington has courageously skipped the 'guided' communication and gone straight to email. I go online to check out this character and find, bizarrely, he looks a lot like a work contact I know who also lives in Paddington. The same contact who has cracked on to me during what I thought was a working lunch (and not a flattering crack-on, the type that leaves you wanting to rock yourself whilst sitting in the shower). Only his name isn't Bob, it's Paul, so it couldn't be him. Or could it?
>
> A second look at his photo and a read of his email confirmed yes, it was indeed 'Paul' from Paddington and not 'Bob'!
>
> Seriously, what is this guy on? What if he actually meets someone, starts dating and it turns out to be serious. Will he wait until she's walking down the aisle to mention that in fact his name isn't Bob, it's Paul?
>
> *(Janice, Australia, dating blog, 2010)*

'Bob'/Paul has not counted on being recognised, but also seems to have failed to recognise his work contact or recall that he had made advances to her during a working lunch. The blog writer can only wonder at how he can expect his deception not to be revealed at some point after meeting. However, there are other occasions where the deception may not be entirely successful, even before parties meet, such as when photographs do not match claims in the text.

Many internet daters seem to diverge from norms of honesty around their age and appearance. One woman 'said she was 28 on her profile', but after 'scrutinising her pictures' a blogger's housemate says, 'If she's 28, I'm about 12' *(Jacob, UK, individual blog, 2013)*. Others comment that a date 'definitely was good looking, but looked a lot older in the flesh—he was also shorter than I imagined' *(Sheila, Australia, dating blog, 2014)*. Steven tells us that one date he meets 'was meant to be about 5 years older than [him] and not 15 [years older]' *(Steven, UK, individual blog, 2019)*. In an earlier post he tells another story of what happens when he meets someone who does not match her profile:

> she looked absolutely nothing like her profile pictures (I think she'd had them professionally done), and if I'm completely honest, as first shallow impressions go, I was a bit disappointed. I feel a bitch for saying this but she was unrecognisable, and although I'm no George Clooney, my profile photographs are always up to date as what's the point in being deceptive about one's appearance when it's going to be the first thing that someone notices (or in my case didn't notice as I didn't know it was her) when you meet them?
>
> *(Steven, UK, individual blog, 2010*—describing a date from 2006*)*

Yet, many women feel they must provide a decent photo to attract interest from men, and for some this may mean a glamour shot (Whitty 2008b: 1714). The stated norm of 'honesty' may well thinly disguise common forms of relatively minor dishonesty and in particular the reiteration of constricting gender stereopic conceptions of women's attractiveness. Women daters are more likely to offer inaccurate visual images in order to be considered of at least possible interest by heterosexual men (Anderson 2016; Cohen 2017). Given that decisions about contacting women are frequently based on photos and less often on a considered assessment of compatibility, this norm may actually function to replicate gender inequities regarding assessing women's value in terms of their appearance. Some dishonesty on the part of heterosexual women daters may in this setting be strategically 'compulsory'. This point is sadly underlined by Bruch and Newman's American study (2018). In their large sample—even if its data has limitations associated with its location in American cities with primarily white participants—women's online prospects sharply dim as they age. In this study, 'men's desirability peaks around 50 and then declines' while women's desirability drops from the time she is 18' (Bruch and Newman 2018: 2). However, despite such pressing imperatives for women, it is important to note that women are also not the only ones to choose more flattering photos, as an Australian woman dater recounts.

> I found a large number of men who lied—about themselves and their circumstances: one man I met was at least thirty kilos heavier and 15 years older than the photos he posted on the site—if indeed the photos were of him (I'm inclined to think it may have been a brother.) . . . I was dismayed about people's inability to be honest.
>
> *(Commenter on Phillip's story, Australia, online newspaper story, 2011)*

These relatively small deviations from the norm are not viewed as seriously as fake accounts, where someone creates a fictitious identity for the purpose of defrauding others (Rege 2009). Fake profiles are usually reported and one blogger reassures would be daters that 'you will see these accounts removed' *(Violet, UK, general blog, 2013)*. Yet, there are other, more everyday ways in which daters themselves tailor their own profiles and filter those of others. Filtering is often talked of with seeming honesty and seen as justified as part of a strategy to find the right person. It is debateable whether this diverges from norms of openmindedness and equality, or follows largely unspoken norms about dating within one's own class. Certainly, class distinctions are evident. For instance, one woman says that a potential partner 'makes moronic grammatical errors' *(Lucy, UK, individual blog, 2018)*. Another account says, 'I double-checked my spelling and grammar, limited the number of exclamation marks I used, and kept my LOLs to myself' *(Phillip, Australia, online newspaper story, 2011)*.

That others recognise this as important is reinforced by a reader of this article who comments, '[c]all me fussy, but if a guy can't spell simple words and the first thing he writes is 'Go Bombers' in his profile, well it just isn't going to work' *(Commenter on Phillip's story, Australia, online newspaper story, 2011)*.

That bad spelling is seen as a marker of working-class status is indicated by the immediate association with writing 'Go Bombers', as the Bombers is the nickname for the Essendon Football Club, a Melbourne Australian Rules club supported largely by working-class men. The commenter is clear that this is not what she is looking for in a man. This response is not surprising in that the article commented on appears in the online version of a broadsheet read largely by the middle class. Several other commenters also say that good spelling was important to them in ways that suggest it is often taken as indication of a lower-class background. Some admit that they 'desperately wanted to filter out people who didn't have the same political ideology as [them] or the same level of education' *(Kali, Australia, individual blog, 2016)*. Level of education, or the lack of it supposedly signalled by an inability to spell, stand as acceptable criteria that people can admit to adopting in searching for partners. In other areas deception requires justification.

Divergence from honesty is thought to be sometimes required; practices do not always fit the ideal. And there is the added complexity that 'while greater disclosure of realistic positive information in a dating profile is successful in attracting others, greater honesty is not' (Hamilton 2016: 149). Moreover, people sometimes believe that they will become the person they present themselves as being before they meet in person. Sometimes individuals may see themselves in ways that others do not, so the 'truth' may be contested. Also internet daters, and particularly women daters, may do things like lowering their age to avoid being filtered out of searches based on fairly arbitrary age brackets or on sexist grounds—that is, heterosexual men's well-established tendency online to prioritise searches for much younger women than themselves (Ellison et al. 2006; Hardey 2008: 1123; Murray 2018; Skopek et al. 2011). Rather than this suggesting that internet dating is straightforwardly a very strategic and instrumental benefit-maximising behaviour by individuals, it may indicate the ongoing reiteration and power of dominant societal and highly gendered norms of appearance and 'attractiveness'. For this reason online divergence from expectations regarding honesty may simultaneously signal ongoing replication of other norms. Divergence from normativity does not fall far from the tree and is strongly associated with 'swivelling' back and forth between social expectations and minor digressions from these expectations which offer no challenge to them (see Thompson 2014: 134).[1]

Transgression: What norms apply to searching for potential partners?

The norms around looking for a partner, may vary but the searching and filtering process can be less individualised than usually suggested and can involve some transgression. In transgressing, internet daters temporarily depart from norms about getting to know each other, but not usually deliberately. There are ways in which the internet may give individuals licence to be fairly direct and swift in becoming intimate, but on general sites there seem to be some expectations of gradually becoming acquainted. Face-to-face encounters are still used as a guide, albeit internet interactions are usually marked as different, often more direct. The value of this directness is pondered by Lucy (using her third-person voice) when looking at the profile of a potential date, Jorge.

> There is, however, one topic on Jorge's profile that gives her cause to hesitate. As well as including standard information like his taste in music and his interests, he talks about sex. A lot. In fact there's a whole section about his preferences, how he considers himself a 'dom', how he's kinky, he likes giving pleasure, how he's 'GGG' (Lucy has to Google that one), and so on.
>
> Normally Lucy would swipe right in horror at such blatant evidence of fuckboyery, but this time she pauses.

> Is it so terrible that he's upfront about sex? she wonders. After all sex is important. It's important to her too, so why not be open about it? And he does so in a very intelligent, matter-of-fact manner that doesn't come across as creepy; his whole profile is simply articulate, relatable, and refreshingly candid.
>
> *(Lucy, UK, dating blog, 2019)*

Although Lucy would normally avoid (by swiping right) men proclaiming such sexual preferences, she questions herself and her imagined readers, reflexively considering whether openness about sex might be an important thing to communicate. Her consideration of this perceived transgression of norms about not sharing sexual information too soon, is done in interaction with imagined, generalised others.

Searching for partners can also transgress norms encouraging individual searches for 'the one'. Accounts give various examples of daters helping others in their search for a partner. One such instance is explained by a blogger.

> I have been on [this dating site] for a while and came across a profile of a very nice guy. I didn't think he was suitable for me but I had a friend in mind that I thought he would be perfect for and contacted him asking if I could set up him. Yes he must have thought it was mega strange a single girl asking him to go out with her friend. However they did and now 2 years on, they're still together (living together in fact!), very much in love and I'm insanely smug about the whole thing.
>
> *(Sula, UK, dating site blog, no date)*

Another dating site blogger tells how he befriended a man he met at an event organised by a dating website company in 2014 and 'two days later . . . helped him out with a double date through a girl he had been in contact with on the website!' *(Robin, UK, dating site blog, no date)*. Other sites also tell of members such as a woman who 'made friends with another member on [sic] one of our events. Then she was set up with her brother' *(Anonymous, dating site blog, UK, 2019)*. Another tale tells of 'a man and woman who had met on match and been on a few dates and, while they really got on with each other, they didn't fancy each other and so they had decided to come to the event together, to see who else they could meet' *(Fiona, UK, dating site blog, 2015)*. These examples are from blogs sponsored by dating sites, which it might be assumed are designed to help 'sell' the sites as places to find their perfect match for a long-term relationship. It is also in the interests of the sites to make people aware of the unexpectedly useful consequence of making friends when these assist the more explicit goal of matching life-partners. However, elsewhere in the stories there are examples of people making friendships rather than romantic relationships via internet dating sites. They say after meeting someone for a date: 'we decided to stay friends' *(Steven, UK, individual blog, 2010)* or to 'stay in touch as friends' *(Jacob, UK, individual blog, 2013)*. There are examples of this in the vast majority of dating accounts.

Yet, some caution is required, because in many cases, agreeing to be friends seems to be a new norm, a polite way to disengage from 'unsuccessful' dates. Many daters reported going on a date and then one party asking if they could 'be just friends' but then they 'never saw or heard from her again' *(Steven, UK, individual blog, 2011)*. It is not clear how often friendships with people once dated are actually sustained. The slippage from dating into friendship may not involve deliberate or lasting efforts to upset norms that centre around romantic coupling. Nevertheless, the development of friendship through internet dating, particularly when intentionally sought, offers intriguing possibilities. There is certainly room for further research on friendships made through online dating. This topic is discussed in more detail in Chapter 5. For now we turn to other more clear-cut subversive examples.

Subversion: Who initiates? Who makes first contact?

Although previously in the book we noted that many women do wait or still feel that they have to wait for messages from men (see Chapter 2), there is some indication that these norms are subverted. Subversions are more deliberate than transgressions, but may remain rather quiet disruptions, for example women initiating contact (see also Hardey 2008: 1125), rather than the norm of men taking the lead. For instance, in commenting on an article about internet dating one man says, 'I used [dating site] for a number of years before finding my lovely wife (actually, she found me). I still met and dated women in the traditional way, online introductions complimented the approach' *(Commenter on Phillip's story, Australia, online newspaper story, 2011)*.

UK men also report instances of where '[s]he made contact with me first' *(Steven, UK, individual blog, 2015)*, although one UK woman blogger is a little surprised when she checks her phone and sees that she made the first move in contacting a date *(Lucy, UK, individual blog, 2019)*. In recent years, women making first contact does still seem to be unusual enough to cause surprise or be worth remarking on, as another blogger reveals,

> [w]ithin a few minutes of logging on [to a Jewish dating site], I had started to get messages from quite a few stunningly attractive women, most of whom were moderate British Jews of the same sort of stripe as myself. I can't decide whether there is an imbalance of men to women, or whether Jewish ladies are just more forthright, but either way, it's great.
>
> *(Jacob, UK, individual blog, 2013)*

The experience of being approached by women in 2013 still warranted comment and/or attempts at explanation. It continues to give rise to considerable discussion. Many of the men's accounts insist that they would appreciate and respond very favourably to women initiating contact and there is evidence that women who initiate are more likely to get a response than men who do the same (Abgarian 2018; Toglia 2016; Victor 2016). However, there are also indications that, while men do respond to women contacting them, what these men appreciate is likely to be the opportunity for greater access to women daters rather than necessarily welcoming women sharing in control over contact and relationship progress.

The *Bumble* app was developed in 2014 with the express aim of giving women seemingly greater power over initiating dates. Only women can make the first move on this site. Nevertheless, the difficulties which this Australian woman dater experienced on *Bumble* reveal ongoing antagonism to women iniating contact.

> The discrepancy between Bumble's sunny narrative and my stormier encounters stemmed from the app's outdated brand of feminism. The women-taking-charge-for-themselves model assumes that we live in a girl-power bubble. It ignores men's feelings about adopting a more passive dating role. This creates tensions between users. I learned the hard way that despite our feminist advances, many men are still not comfortable waiting to be asked out.
>
> Some Bumble men view the app's signature design as a way for women to rob them of their rightful dating power. Many openly critiqued us for acting 'like men' and I was ghosted, sexually degraded and subjected to violent language by men who resented me or what I represented as a feminist. This was confirmed by several of my matches, who discussed women's acquisition of socio-economic and sexual power as a problem.
>
> *(Treena, Australia, online article, 2019)*

Despite this critique it is evident that different platforms have different norms, with some affording much less capacity for 'forthrightness' from women than *Bumble*. The British Jewish man cited above notes in another post that from discussions with the date he met through *Ashley Madison*, men seem to do the approaching on that site, although he describes his date as 'very good looking' and thus as having to deal with 'a huge amount of messages that were explicitly sexual'. As he wryly comments, his own success in approaching women can be explained by the fact that 'in the world of *Ashley Madison*, if you don't instantly segue to asking to see a woman's breasts in text speak, you are quite the player' *(Jacob, UK, individual blog, 2013)*. The nature of *Ashley Madison* as a site to facilitate extramarital affairs may require women to be more reserved in order to avoid more aggressively sexual approaches from men. These remain frequent on the internet, but, as we noted in Chapter 2, such approaches may be easier for women to deal with online (Hardey 2008:

1115–1118; 1125). As another woman comments, 'it is much easier to ignore someone online than get rid of someone who is chasing you all over the pub' *(Carol, dating site blog, UK)*.

Despite these instances, change is rather patchy and norms of men pursuing women are still regularly invoked by those in our sample. For example, one internet dater reports on a failed relationship, saying that he 'later found out that she'd stopped contacting [him] because [he] hadn't actually asked her out early enough' *(Steven, UK, individual blog, 2010)*. And while this illustrates the ongoing potency of gender stereotypes with regard to intimacy,[2] it also highlights that norms around the first date—that is, at what point to meet in person—are still somewhat unclear.

Into heterodoxy? First meetings and who pays (out)

The heterosexual internet dating accounts we analysed provide few examples of more radical heterodoxical practices. Instances of dissidence,when they appear, tend to be rather unassuming rather than openly confronting, and heresy does not seem to be in evidence, although the degree of shift in norms is open to further discussion. For the most part, meeting in person relatively soon is generally advised and from there daters often return to the usual rules for romantic encounters, which goes some way to explaining why women are more inclined to delay meeting offline as they have more difficulty filtering and negotiating men's behaviour once the usual (offline) rules return. It is interestingly in this context that meeting early on is almost always advised, supposedly in order to avoid raising unrealistic hopes. As one Australian internet dater advises,

> [d]on't get your hopes up about any 'date', but if you're comfortable with it, I think meeting in person early on is important. Like someone else mentioned, I too 'met' someone and spent hours and hours talking on the phone with him, but when we met in person we had nothing at all to say to each other.
>
> That said, I didn't respond to the one guy whose first message said 'I don't do this whole email thing. Let's meet.' A little getting-to-know-you email isn't a bad thing!
>
> *(Commenter on Phillip's story, Australia, online newspaper story, 2011)*

Why longer communication by email or telephone 'builds expectations that are rarely met' *(Commenter on Phillip's story, Australia, online newspaper story, 2011)* is typically not explained.[3] This may highlight the importance of the promise or memory of physical presence in building strong intimate relationships (Bauman 2003: 103). However, it is also possible that such advice reflects men's preference for shifting offline quickly and the advice is intended to forestall the possibility that male daters may be deterred by longer online discussion.

In this setting, online or other interactions which defer face-to-face meetings are widely regarded as unusual, often as problematic, but might instead be reconceived—given the strong advice and indeed pressure to meet offline quickly—as transgressive or even perhaps subversive practices. Most of the first offline dates describe meeting for a drink, a coffee or a meal. Typically, the norm appears to be to meet in a public place of some kind, so that there is less risk involved—particularly less risk of harassment or violence for women. This form of meeting is repeatedly recommended in the advice many sites give about the first date. The advice they give also suggests letting friends know where you are going *(Anonymous, UK, dating site blog, 2014*; *Kali, woman, Australia, individual blog, 2016)*. The lack of advice daters in our sample seem to give each other about first dates may also arise because dating and other online sites provide considerable guidance on this topic. Some sites also provide events for daters to meet each other and bloggers write about these events, perhaps having been encouraged to do so by the promise of free access to the site or other rewards. Thus, these should be read a little like advertisements, even if they still have things to tell us. Daters may approach these organised events differently to dates with individuals. They may be 'perhaps hunting for a boyfriend for winter, perhaps just curious' and thus have 'grabbed a few friends and headed down to one of [the dating site's] events' *(Kat, UK, dating site blog, no date)*. The tendency to go with friends to these gatherings of singles marks out such events as perhaps less bound by usual dating rules and hence as a potentially less normative, more divergent practice compared to the hefty weight of established heteronormative traditions which surround the one-to-one intimacy of a first date with a potential partner. Henriksson's detailed account (2014; 2019) of the variety of organised singles events in Sweden suggests however that views of these group events as compared to individual first dates are likely to be culturally specific and may also reflect age-specific norms within different cultural contexts (see also Li et al. 2020).

Dating sites' advice about first dates tends very often to reinforce heteronormative conventions about gender and its intersections with class, rather than encouraging divergence or more heterodoxical possibilities. For example, one site advises that women 'don't over accessorize or show too much flesh' and men 'be presentable, this means shiny shoes, ironed shirts, tidy hair' *(Anonymous, UK, dating site blog, 2014)*. This evokes prejudices about 'undesirable' working-class women showing too much flesh and jewellery (Skeggs 2005) and appears to have in mind a rather middle-class, shiny-shoed and ironed-shirt version of masculinity. The site writers appear not to see the contradiction in advising those who may not normally dress like this that 'clothes should be fuss free, clean cut, and should say something about who you are at the same time' *(Anonymous, UK, dating site blog, 2014)*. A careful management of one's appearance and a degree of restraint is in fact being suggested and dominant middle-class norms about appropriate gendered forms of presentation reign in a way Elias (2000) would recognise. This inclination to buttress gendered heteronormative conventions may well not only be found in site advice but additionally in the more disguised and embedded processes of matchmaking itself. Kelley and Malouf (2013: 5, 7) noted in their study of newspaper matchmaking columns that while gender-differentiated preferences regarding age are commonly asserted by online daters—along the lines, for example, of men preferring younger women—these asymmetrical relations may well also be maintained by matchmaker bias in favour of age disparity (see Chapter 6 for further discussion). This intriguing research has implications for online dating sites since such sites vary considerably in the degree to which they shape and shepherd partner selection (Kreager et al. 2014). Mearns (2020: 5, 7), for example, points out that identity, sexual and attractiveness/bodily hierarchies remain stark in internet dating and are maintained by the dating sites themselves. Most sites still require daters to be given 'male' or 'female' assignment and biases regarding 'expectations and norms of power' arise in most dating services, including within algorithms like those used in *OkCupid* which sort and match bodies such that those designated as 'attractive' are seen only by others deemed similarly prepossessing.

Yet, despite these pressures towards orthodoxy, some subversion of gendered heteronorms is apparent, if far from endemic within internet dating, around who pays for the date. There are numerous examples in these accounts of internet dating of remaining gender expectations that men pay for dates, as one man explains,

> I too have had mixed experiences [of internet dating] and am still trying. I tell you what, it does challenger ones perceptions of women what all of us are capable of [sic]. Do you think when women say, 'I enjoy weekends away' that is a subtle hint she will give you want [sic] you want, provided you pay and come up with the goods whatever that may be? Or is it just another 'throw away line'? I'm sure women have their frustrations too. We could go on and on.
>
> *(Commenter on Phillip's story, Australia, online newspaper story, 2011)*

This man wonders whether the promise of men getting what they want (presumably sex) from weekends away is in fact only obtained if men do what is expected and pay. Another man says 'she took me to a really swanky place (I paid, of course)' *(Peter, individual blog, UK, 2005)*, with the 'of course' highlighting the taken-for-granted nature of such gendered expectations. These traditional expectations are also evident in the women's accounts. For example, a woman blogger mentions a date's 'reluctance to open his wallet' even if in this case the woman claims she has a 'must-be-an-exceptionally-independent-woman-and-pay-my-way-in-a-somewhat-confrontational-manner-thing'

(Rose, Australia, dating site blog, 2013). More recent blogs suggest that there is not a set norm about paying. Lucy *(UK, individual blog, 2019)* indicates this when she reports on a date where 'to avoid any polite confusion over who's paying, she firmly puts her card down on the tray and says, with the same fake, cheery tone, "We'll split this, yeah?"'. Divergence from the heteronorm is patchy, and Lucy's fake cheeriness suggests some lingering expectation that men should pay. These examples hint at only a slight relaxation of the tenacious hold of gendered dating discourses and practices that reflect gender stereotypic models of women's passivity and financial dependency on men. There is not much in the way of evidence of the emergence of new norms around this issue. All the same, it is worth mentioning here that while women in many countries—for example, USA and China (Hitsch et al. 2010; McWilliams and Barrett 2012; Su and Hu 2019)—appear to be guided in their dating preferences by men's financial/occupational status, there are nevertheless some recent signs that concern with income is lessening for women (and men; Dinh et al. 2018).

We now move on to a range of perhaps even more more murky issues where there is very little or decidedly equivocal evidence of the development of new norms in relation to gendered power relations. First in terms of these issues is perhaps unsurprisingly, sex. Sex follows on from our deliberations about payment, since the question of who pays may well be pertinent to sexual interactions. The norms about *when* to have sex appear to be at least if not more complex than the issue of paying and are seemingly variable rather than inevitably following well-established normative paths.

How heterodoxical internet dating relationships are in sexual terms will of course vary considerably, but how accurate are claims that having and 'talking about sex right from the off is completely the norm' in internet dating *(Lucy, UK, individual blog, 2019)*? Australian research on women internet daters reveals that 41.5 per cent of these women met new sexual partners online, but perhaps they were hoping for something lasting as over 60 per cent claimed they were looking for a long-term partner (Bateson et al. 2012). Data from the US suggests that heterosexual men are more likely to use online dating to find casual sex partners than women (Gatter and Hodkinson, 2016). However, both sexes may disguise that they are interested in 'hook-ups' because they think this will not be well perceived (Zytko et al., 2018: 69). This implies that direct use of internet dating for causal sex is not normative, even though it may in fact be relatively common. Normative—perhaps particularly in the realm of intimacy and sexuality—should not be considered equivalent to usual; it may indeed be aspirational or amount to a claim to respectability. That said, the overall point regarding the somewhat partial, limited and inconsistent social acceptance of casual sex in heterosexual internet dating differs between age groups and is certainly shaped by different cultural contexts. For example, casual sex, though available through particular websites, remains not widely accepted as an aim in China. Most heterosexual dating sites in China emphasise marriage and references to casual sex may also result in government censorship and website warnings (Li and Lipscomb 2017).

Where sites or apps are specifically oriented more to facilitating casual hook-ups, it is not surprising that sex may occur as part of the first meeting, but in other cases norms around the timing of first intercourse tend to favour waiting (Hardey 2008: 1124–1125). This may be true even for dating apps like Tinder, where evidence does not support common perceptions that it is just used to facilitate casual sexual encounters. However young daters are more likely to use it in that way (Gatter and Hodkinson, 2016). Other sites do seem to have a clear aim of providing sexual hook-ups and some are directed at older women. One blogger describes a 'Cougar' dating site he explores as 'very clearly a fairly sex-oriented site—lots of the pictures are overtly sexual, with stockings, suspenders, and low cut tops a particularly common trope' *(Jacob, UK, individual blog, 2013)* (see also Chapter 6). Occasionally women daters document casual sexual encounters, like a woman who describes contacting a man via a dating app and going to his house because she 'hadn't had sex in 10 months, and [she] was in desperate need for some action' *(Chiara, Australia, guest post on a dating blog, 2019)*. The accounts of women in our sample generally reveal little about women using sites for casual sex or about how they might navigate the timing of sex with partners they meet online. Nevertheless, the existence of so-called adult online dating sites such as *Adult Friend Finder* demonstrates at the very least that some women do engage in short-term sexual encounters (Escasa-Dorne and Jankowiak 2018).

Sex on the first date, whether it leads to a relationship or not, seems somewhat transgressive; it may occur but is not often admitted to, especially by women daters. Lucy *(UK, individual blog, 2019)* does blog about the sexual partners she has met online. Yet, apart from one man she was already friends with, she says she does not have sex with them until, at earliest, the second date. It is difficult to know how 'normal' this is. Scarcity of blog accounts by women internet daters (although see Muise 2011 for an account of women who blog about sex) about their casual sexual adventures is not suprising given unwanted attention this may get them and the continuance of sexual double-standards that judge 'promiscuous' women harshly (England and Bearak 2014). There is nonetheless the occasional glimpse of women presenting themselves as open to casual sex, as in the case of one blogger's date who describes herself as 'the copping off with someone in a club type' *(Jacob, UK, individual blog, 2013)*. Dating sites such as *Local Slags* also appear to present women as open to sex for fun. However, as we have already noted, knowledge of women's sexual experiences as internet daters is limited, beyond some broad claims that online daters usually take between three to six dates before having sex, are increasingly not having sex on their first date and women daters are less interested in first-night sex than men (see for example, Holohan 2016).

Despite this rather limited knowledge base, it appears that online dating norms regarding sex still show gender differences, even as these might be changing. Many academic researchers, in-house surveys, commentators and internet daters reiterate that women are slower to register interest in sex and take longer to arrive at the first sexual encounter. There is a long history to this gender differientated account of heterosexuals with a number of explanations attached. Given the differential risks of dating this would hardly be surprising, quite apart from the continuing complexities associated with women declaring sexual agency. We would add here a contributing issue that is almost never raised in the literature—that is, the ongoing common understanding of hetero-sex as equivalent to penis–vagina intercourse. This remarkably tenacious 'coital imperative' (Jackson 1984; Willis et al. 2018) may well negatively affect heterosexual women's assessment and experience of the pleasure 'pay-off'' of sex while offering comparatively disproportionate gratification to heterosexual men (Brewer and Hendrie 2011; Cherkasskaya and Rosario 2019; Compton 2019; Hayfield and Clarke 2012; Mintz 2015; Rowland 2020).[4] Yet, women do also sometimes acknowledge their delight in finding a way of meeting men that enables sex. For example, 84-year-old Giselle was astonished after more than a decade of no sexual intimacy to find that she could begin again: '[w]ell, I wouldn't have thought it was possible to go there at my age . . . We'd both been starving but never really expected to have a sexual relationship again' (*Giselle, Australia*—cited in Arndt 2016).

While a gendered differential in relation to sexual interest is largely assumed in the available literature and by many participating in internet dating, this does not mean that declaring sexual interest is straightforward for men. In the case of more 'standard' sites where people are seeking longer-term relationships, even male daters seem to feel the need to point out that they do not *usually* have sex on the first date.

> As it turned out we got on really well, both got pretty hammered, and for only the second time on a first date, we got a cab back to hers, where she made me a halloumi sandwich (not a euphemism) and we had sex. Considering I'd been experiencing a personal drought for almost two years, I was amazed as to how well things went.
>
> *(Steven, UK, individual blog, 2013)*

This prolific internet dater claims he has only twice had sex on a first date. Yet, there are many stories of sleeping together on a second date. For example, Steven describes another woman, saying they 'met for our second date 5 days later and slept together after a museum visit and dinner near where she lived' *(Steven, UK, individual blog, 2015)*. Another account from a UK man describes how 'a couple of weeks later [after the first date] when I was back in the UK. . . . We went out around [her town] and she started telling me some of her intimate stories . . . I stayed over and in the morning we talked about when to next get together *(Peter, UK, individual blog, 2005)*. Thus the accounts of the men in our sample do not seem to show them using internet dating just to pursue casual sex and other research confirms that only some men are 'players' (Hardey 2008; see also Bean 2011: 175–209). Those men writing a blog about their internet dating experiences might be atypical but other methods such as interviews are not necessarily any better at getting to the 'truth' about men's experiences of using internet dating to engage in casual sex because they rely on self report and men might be more likely to boast about (sometimes non-existent) sexual conquests (Bean 2011: 175–209). Norms around 'waiting' before having sex with partners may not be dramatically altered by online dating but it can raise new issues around monogamy.

When do you become exclusive and take your online profile down? Norms of monogamy and coupledom

There is not a lot said in these accounts about the uncertainty around at what point a 'serious' relationship begins, but there are glimmers of subversions around monogamy. Introducing the partner to friends and kin seems one marker, also common to relationships started offline (c.f. Hardey 2008: 1123). One blogger related the shift from trips to the 'theatre, meals out, the cinema, picnics, her meeting my friends, me meeting her friends, us deciding we were a couple, and indeed, last week, her meeting my mum. Yes, it's proper serious and everything' *(Jacob, UK, individual dating blog, 2013)*. There are other examples where one partner thinks a relationship serious, but then she 'let slip that she had someone else who she saw off and on when they both had free time' *(Steven, UK, individual dating blog, 2015)*. The seeing of others is taken to show she is not 'serious'. Yet, it remains difficult to pinpoint what leads to the daters deciding they 'were a couple'. Some insight appears however, in one woman blogger's musings on the topic when her internet dating relationship becomes more serious. She asks,

> When do we become 'exclusive'?
>
> When do we have the chat about going offline?
>
> . . .
>
> I'm still online.
>
> So is he.
>
> I've not been remotely interested in seeing anyone other than him and I have a feeling it's mutual. He calls me up for chats, we meet at the weekend, I know what he's doing almost every night this week . . . But he COULD still be seeing other women, writing messages or sending virtual kisses.
>
> Is it likely? Rationally—no: he's a busy man. But my personal confidence-dementor is sucking out the happiness and shovelling in doubt.
>
> At what point do you reject the thousands of potential 'others' to concentrate your attentions on just one?

> In my head I've already gone down the exclusive/offline route. Consequently and unintentionally I'm starting to behave like A Girlfriend. You know: little hugs, kisses in public, making fun of his sunburn, having arguments with waitresses, letting him see me without makeup, the usual stuff. The stuff you really should keep hidden until a) they've gotten to know you fairly well, b) have introduced you to their friends and c) so besotted it doesn't matter what you do they'd forgive all your foibles.
>
> It's pretty tricky being self-aware yet giddy with hope. It's like trying to drink whilst laughing. There's a fair chance you'll choke.
>
> I'm afraid I'm going to ruin it before it's even started.
>
> *(Rose, Australia, dating site blog, 2012)*

This self-aware assessment highlights the difficulty of navigating the boundary between 'seeing' someone, while perhaps still seeing or being open to meeting others, and becoming 'A Girlfriend' or Boyfriend. Although this issue may arise in all new romantic relationships, the decision about when to reject the multitude of potential 'others' online and focus on only one person is new in both degree and kind. As is the question of when to make the deliberate and semi-public decision to take down one's online profile. Taking down a profile amounts to a declaration of having judged your relationship to be serious and exclusive (Bean 2011: 179–181). Of course offline relationships involve such decisions too, but there is not the same requirement actively to announce this.

As Rose's blog outlined above indicates, while deciding how to assess the seriousness of a sexual relationship is a usual part of emerging romance, online dating clearly involves significant grounds for continuing uncertainty about whether your partner has dropped other options. She is unsure whether he has stopped contacting other women online and expresses anxiety about asking him, in case she might 'ruin it', by seeming too eager. The 'rules' around when to enquire about the possibility of deleting online profiles, let alone actually removing them, are not well established, nor necessarily obviously gendered (although see Bean 2011: 179–181), and appear to rely on emotionally reflexive negotiation between prospective partners. Indeed, the 'issue' of when and how to signal a turn to monogamous intimacy may sometimes result in rather less prescriptively gendered possibilities. Consider for example the instance of an Australian 63-year-old woman who developed new sexual and emotional confidence arising from new experiences of casual sex associated with online dating. She undertook several short-term flings before moving into an onogoing relationship and noted that, '[t]hose experiences are what have enabled me to bring to our relationship what I now do, and he loves every minute of it' (*Lee, Australia*—cited in Arndt 2016).

From our sample it seems that online dating gives rise to a commonly acknowledged period of equivocal instability about how long one can continue to maintain access to and meet others while pursuing a particular relationship. There are normative tensions associated with how long one can maintain an online profile which enables continuing access to potential dates after meeting a person offline a number of times and relatedly how long one can continue to meet more than one person before a decision regarding exclusivity becomes necessary.These widespread tensions about what constitutes appropriate etiquette regarding developing an intimate relationship produce a sense of normative expectations but not a set of clear guidelines about how to respond to them. In consequence such expectations about the proper progress of a romantic/serious relationship may be somewhat destabilised. This uncertain interval, during which daters consider how serious their romance might be, also arises in offline heterosexual relationships, but awareness of other options is typically much more readily available to all online daters. The clear availability of other options can encourage meetings with others for longer online, if only as fall-back 'insurance' in case of the failure of the new relationship. Such instability enables the possibility of transgression from traditional gendered norms, including what we describe as the emergence of 'mundane polyamory' (discussed in more detail in Chapter 4). Decisions about going back online after a relationship ends are perhaps even less subject to social prescription.

When do you go back online after a break up? Devalued or valued connection?

Where serious relationships are formed, they may later founder and come to an end, leaving questions about when to try again that require new norms. These new guides to conduct seem divergent, although rather weakly divergent, from mainstream offline standards of serial monogamy. For example, the rather complicated relationship described by the woman blogger in the previous section goes along for a while, leading to them coming off the dating site at some point, but it all eventually comes to an end.

> So, after a few days of feeling teary (read: blubbering wreck) and a little out of sorts (read: totally rejected, wondering why on earth he hadn't realised that after a 2.5 week break of next to no contact it was always going to be a little awkward) . . . I decided to get back online.
>
> *(Rose, Australia, dating site blog, 2013)*

Here, the relationship has stalled after no contact for a couple of weeks and then it is only a few days until Rose goes back online. It is not entirely clear how long this relationship lasted, and that may have some bearing on how long someone waits before reviving their internet dating profile. A few in the sample describe leaving their profiles up on dating websites for long periods as a means to 'hedge their bets' whatever they deem their relationship status. Some leave their profiles active, if unused, for a short interval when they are in a 'serious' relationship. Still others take their profile down from view (which involves the option of 'taking a break' available on most sites), yet they do not remove their profile from the internet dating website and terminate their account. And, it would also seem that 'expert opinion' on this issue is variable, signalling a rather loose or at least less self-evident set of social norms about 'moving on' (Deitz 2016).

Whatever the case, our sample daters diverge in terms of a shorter period of waiting between relationships; they are for the most part ready to be called rapidly back into service. Or they return, maybe a few months later to a different site, wanting to 'get back on the dating horse again' *(Steven, UK, individual blog, 2015)*. This suggests some individualisation and might seem to confirm Giddens's (1992) portrayal of the modern triumph of 'pure' relationships, which people exit as soon as they are not good and then seek a new special relationship. There may be elements of this in the practices described, but internet dating is far more about connecting with a range of others than such portrayals suggest. Moreover, Rose's experience cited above, tends to highlight the complexities, sorrow and pain of negotiating one's intense desire for connection and heteronormative coupledom rather than illustrating the behaviour of a seamless, clear-headed, autonomous, rational actor. While the social guidelines regarding overcoming loss and returning to seek romance may not be set in normative concrete, they are evidently not purely individualised either. At minimum, online dating practices exist in the setting of a relatively new technology/business of intimacy which allows immediate access to new potential partners—that is, they are subject to profit-making technological innovations precisely enabling repeated romantic searches. Moreover, this technological access arises in a context of compelling social pressures to resume the search for heterosexual romance that Adrienne Rich (1980) once famously described in terms of 'compulsory heterosexuality' (see also Richardson 1996; G. Seidman 2019; S. Seidman 2009). Individual daters are hardly atomistic 'lone wolves' floating free of material socio-political imperatives.

Do new norms indicate a shift in gendered power relations?

By looking at internet dating etiquette we can see how norms alter in relation to socio-historical processes and shifts in power relations. There is evidence here of the shoring up of class and gender hierarchies, but also of them being undermined to some degree. As the book progresses we will see in more detail how this works. What is evident here is the relational and even ethical ways in which internet daters deliberate and act.

Norms of internet dating are relationally, collectively, enquired about and evolve in a participatory way as power and status differences become a little less immovable. There may be tensions and disagreements and contradictions, but there is considerable individually expressed and shared ethical concern about what kinds of behaviour are appropriate. We see divergence from heteronorms in tensions between norms of truthfulness and of creating a good impression and the pressure women still face to be physically attractive in stereotypically feminine ways. There is also evidence of transgression of norms about dating as an individualised search for one romantic partner in attempts to help others find love and to disengage gently when a date is not what was hoped for. Moreover, the women in the sample accounts, and in other sources, indicate some subversion of gendered heteronorms in that they have some opportunity at least to choose men rather than wait to be chosen, and they may exercise some control by delaying when to meet in person. Some advice on looking for 'good' clothes or spelling appears to reinforce class distinctions, and certainly old gendered norms about men paying for dates linger. At the same time, in our sample it seems further subversive possibilities arise around some women being able to seek more sexual encounters while many of the men seem to seek to put less emphasis on sex. When these internet daters do meet someone they like, the decision about becoming 'serious' may diverge only a little from serial monogamy, but has to take heed of new worries about whether and when that means taking down your dating profile.

The management of norms amongst these internet daters seems to vary based on the strength of ties involved—or hoped for (as Greti Ivana 2018 found in relation to Facebook). The accounts show processes of social innovation at work and how norms are less dictated by experts/authorities than in the past. Power is still at work, but it is somewhat more diffuse. Though power relations clearly flow through a range of normative discourses (such as discourses about masculinity as taking charge in relation to relationship progression, sex and money), power also depends on concrete markers of class distinction in terms of things like who has most time or money to engage in internet dating. Furthermore, those with more technical expertise are equipped with a form of cultural capital which may enable them to be more likely to control when and how 'discussions' of norms occur. Still, this chapter focuses on internet daters who have good digital capabilities given that many are able to produce blogs or posts, not just navigate internet dating sites in a wobbly fashion. Yet, even lesser skills are sufficient to enable potential participation in the continual construction and reconstruction of social norms online. Here, we have begun to explore those processes through daters' online accounts of their experiences.

The results suggest that as power relations, especially around gender, shift, then heterosexual internet dating etiquette reveals a greater need for individual, intersubjective and social reflexivity in deciding how to act. Negotiating this evolving etiquette involves an interactive labile dance—what might be described as the play of a 'micro-politics' of social innovation—through which some possibilities for change can emerge (Beasley 2015; Beasley et al. 2012: 88).[5]

Notes

1 We make use here of the term, swivelling, coined by Tulia Thompson (2014) in response to related terminologies developed by Sarah Ahmed in her book, *Queer Phenomenology* (2006). Ahmed outlines how departures from heteronormativity are 'straightened' up, or even 'stopped'. Thompson adds to this vocabulary by noting that various efforts to to do gender and sexuality differently often involve swivelling between non-normative and normative practices rather than any simple uninterrupted opposition to heteronormativity.

2 We note here that a number of studies on this pattern of men initiating and women waiting with regard to intimate/sexual connections take as given that it demonstrates and confirms innate sex differences in 'mating choices' derived from evolutionary reproductive strategies (see for example, Abramova et al. 2016; Trivers 1972; Buss and Schmitt 1993). Suffice to say that gender, feminist, queer and critical masculinities scholarship, along with a range of other approaches, generally dispute the self-evidence of this biologistic/biological determinist framework (including DeLameter and Hyde 1998; Fehr 2011; Kimmel 2016).

3 Hamilton (2016) suggests that the swiftly 'hyperpersonal' character of online dating exchanges can lead to idealisation of the dating partner, but unusually she notes research to support this view. Ramirez et al. (2015) found that online communications had a timing tipping point (between 7–23 days) whereafter meeting face-to-face was likely to be 'more disappointing than successful'. On this basis Hamilton recommends internet daters move to meeting offline 'within a reasonably short period' (Hamilton 2016: 151).

4 The marked inclination of many if not most sources is to presume that heterosexual men's supposedly greater interest in sex than women is derived simply from being male and to ignore the ways in which their sexual interest and desire is continuously affirmed, indeed demanded, such that it is deemed a compulsory feature of masculinity. This 'male sex drive discourse' means that men who do not feel especially sexual would be hard pressed to admit it (Cherkasskaya and Rosario 2019; Garcia-Favaro 2016; Gupta 2015; Hollway 1989; Thompson 2018).

5 In referring to the micro-politics of internet dating, as we briefly outlined in the Introduction, we mean here to draw attention to a broadened conception of 'the political', of power relations, which links private innovations and larger-scale developments.

Chapter 4

Emotionally exciting novel heterosexual practices?

Internet dating is becoming a major way in which intimate relationships develop, but the emotional experiences involved in internet dating, where touched upon at all, have largely been considered in terms of fear of risk (Barraket and Henry-Waring 2008; Hardey 2002; Whitty 2008a). Instead we focus on internet dating as a potentially exciting and innovative form of heterosexual relating. Studying internet dating provides opportunities to appreciate the sociality of embodied flesh and the importance of emotions in making, shaping and changing heterosexual relations to others. Heterosexuality is an embodied, relational and emotional matter, even if it begins on a screen. Indeed some authors have claimed that online relationships can be more radical because of the lack of social cues, the chance to try out new identities, to overcome inhibitions or to transgress gender roles (Hillier and Harrison 2007; Lea and Spears 1995; Turkle 1995). More recently, researchers tend for the most part to be more sceptical about whether relating online is a space where people could be free from the constraints of gendered and sexual subject positions and practices. Attention to risk is now more likely to predominate (Baym 2015:96). In this setting, while we acknowledge that, for example, user narratives around internet dating can involve a normative emphasis upon risk, we also insist that this is by no means the whole story. Thus, we outlined in Chapter 2 that the prevalent nutter narrative is perhaps primarily to be understood as expressing fear of risk, but at the same time we pointed out that even this story is not entirely without opportunities for social change. In this chapter, we further resist any singular concern with the dangers of dating. Questions remain about the extent to which internet dating might enable exciting forms of heterosexuality that challenge gender and other inequalities. Rather than focusing upon fear of risk, this chapter examines how other emotions, especially the excitement involved in internet dating might allow for some disruption of gendered heteronormativity.

Despite awareness that online dating has particular features (such as an intensified feedback loop produced through the more deliberative composition of online conversations and online disinhibition) which intensify emotional interactions (Hamilton 2016: 151), literature on internet dating has paid strangely little attention to the emotions involved, let alone to its potential for producing transformative excitement. Attention has centred on a sense of threat, with anxiety, lack of trust and fear presumed to be the preeminent feelings (Baym 2015: 96), along with some attention to other emotions such as distrust and disappointment in discussions of internet dating as a new form of consumption (Barraket and Henry-Waring 2008; Heino et al. 2010). However, this account of dating as a risk assessment exercise concerning consumption choices amounts to an understanding of internet dating as about individual rational calculation, selling the self, and 'shopping' for love/sex/partners (for example, DeMasi 2011; Heino et al. 2010; Jagger 2001). As we have mentioned in previous chapters, particularly in Chapter 3, there is continual reference to internet dating as risky with regard to *representation*—in encouraging a commodified objectification of self and others, and to *misrepresentation*—with calculating use of deception depicted as rife (Barraket and Henry-Waring 2008; Baym 2015: 96; Heino et al. 2010). We wish to dispute this cognitive, strategic model of the self and social interconnection. A concentration on the commercialisation of intimacy (Hochschild 2003) or commodification of emotions (Illouz 2009, 2006), often linked to bemoaning increasing 'inauthenticity' or alienation from human sociality (Miller et al. 2016: x, xiii, 100–113), is not necessarily the only or best way to understand internet dating.

A few writers have at least raised the issue of whether internet dating enables sexual and social innovations or simply promotes the continuation of existing modes of meeting and matchmaking (Barraket and Henry-Waring 2008: 149; Finkel et al. 2012; Heino et al. 2010: 428; McWilliams and Barrett 2012). Yet, even in these analyses, the potential of internet dating is very often narrowly conceived as lying in the new technical options it provides such as making more potential partners accessible or allowing people to use algorithmic technology to find a partner with 'matching' characteristics (Baym 2015: 113), in spite of the arguable effectiveness of such algorithms (Finkel et al. 2012; Hamilton 2016: 154). What is frequently lacking is any questioning of whether either safety or risk might feel something other than scary and whether these feelings might prompt or discourage new ways of connecting that rework power relations. In particular, we look at possible effects upon gendered heteronorms arising through internet dating.

Our intention is not to return to the digital idealism of early researchers like Donna Haraway, with her emphasis upon the transformative possibilities of technology to reconstitute gender subjectivities and gender power relations (Haraway 1991; see also associated debates around cyberfeminism including Castells 1997; Daniels 2009; Scott 2016; Shade 2002; Wilding 1998). Rather, in keeping with the contextual orientation towards social media outlined by Miller et al. (2016: 114–115), among others, our study suggests a complex picture of social change with regard to gendered heterosexuality. In this chapter we outline that complexity—what we described in the last chapter as the micro-politics of social change—by undertaking a less common approach to technologies of intimacy and its impacts. Our approach foregrounds embodied emotionality rather than the cognitive rational individual at the centre of much contemporary research on digital sociality.

Some conceptual invention is required to understand social and sexual connections made via the internet in more emotional and embodied ways. Beasley and Bacchi speak of 'social flesh' to make sense of care, sexuality, citizenship and sociality *per se* in ways that highlight the significance of human interdependence (2012, 2007, 2000). In addition, it can be helpful to consider the emotional reflexivity through which intimate links with others are formed and maintained. Emotional reflexivity is about reflecting on and enacting interpretation of emotions, in interaction (Holmes 2015, 2010). This chapter initially uses these concepts and builds upon our previous theoretical work on heterosexuality to provide a basis on which to consider whether internet dating is exciting or encourages other novel emotions.

First, however, we counter representations of internet dating as about highly rationalised and commodified selection of partners requiring management of risk. Experiences of using internet dating to find relationships may in some ways be like shopping, but arguably in more enjoyable, sensory, embodied and emotional ways.[1] We then examine whether internet dating excites forms of heterosexual practice that might wander from, or even undermine, aspects of heteronormativity. We consider how emotional reflexivity can, under some circumstances, transform practices of intimately relating to others. The chapter disputes theoretical and empirical accounts of risk as feared by offering an alternative view of internet dating as exciting. It suggests that risk is not equivalent to fear or negative withdrawal from dating. Rather, we note that other emotions which may or may not be linked to risk assessment, such as excitement, can prompt an emotional reflexivity capable of interrupting or even upsetting hierarchical relations to others. Thus, as in earlier chapters, we give examples of how heterosexual internet dating practices can sometimes *diverge from*, *transgress* and *subvert* the heteronorm. Here, the analysis suggests that internet dating can, for example, emotionally excite heterosexuals to move beyond and undermine (inequitable) homogamy and monogamy, albeit not inevitably in radical ways.

Theoretically exciting heterosexual internet dating

This chapter addresses two gaps identified in the existing literature on internet dating: a gap around emotions and a gap around directions for liberating forms of risk. In the Introduction (Chapter 1) we outlined our approach to heterosexuality, internet dating and emotions as founded on theoretical work about innovations in heterosexuality (Beasley et al. 2012). Through an understanding of heterodox (non-normative) heterosexualities we explore not only what kinds of emotions might be involved in internet dating for heterosexuals but how the emotional experiences might be related to positive shifts in hetero-gendered sexualities. We do not accept views of heterosexuality as monolithic and as inevitably heteronormative (see Warner 1999). To suggest that heterosexuality is not always heteronormative challenges the hegemonic coherence of (gendered) heteronormativity and thus avoids seeing social change as only happening at the social margins. This helps to 'trouble' heterosexuality, to undo its illusory homogeneity and authority as occurs in Butler's 'undoing' of gender (Butler 2004). It also makes possible some understanding of how such undoing might be informed by emotional reflexivity.

The challenge we undertake is to understand internet dating not simply as a matter of managing fear of risk but in terms of emotionally reflexive practices arising from and contributing to the ongoing construction of selves. This approach to internet dating may be understood in a range of ways. Ulrick Beck (1992) has argued that reflexivity has arisen as a response to the expansion of risk associated with the detraditionalisation and individualisation associated with modernity. As 'traditional' norms and practices no longer provide a clear template for people's actions, he maintains that they are forced into reflexive self-confrontation in deciding how to live and relate and must calculate the risks involved in various options. We suggest that Beck's characterisation of reflexivity remains marked by an understanding of modern subjects as rational agents compelled to reflect and act as individuals. The result is a tendency to constitute the self as separate from social structures and as a unified and stable platform for action that can resist social prescription, located outside the self. In the process of characterising subjects as distinct from the social context in which they become subjects, Beck also offers a thin account of subjectivity which not only rests on a notion of subjects as pre-existent (though historically specified) unitary agents but additionally narrows this conception of subjectivity to one that is fundamentally cognitive. He underestimates the more relational, reactive, less conscious, less calculating and more visceral, sensate aspects of embodied subjectivity.

Where it does allude to emotion, Beck's account of modernity as 'risk society', overemphasises fear and fails to substantively attend to a wider complex of emotions that might be associated with risk and possible responses to it, including emotional reflexivity. In particular, his account is not inclined to register the possibility of an engagement with, or even embracing of, risk and other associated positive responses such as enjoyment, excitement and pleasure. Beck's approach remains mired in his emphasis upon a cerebral, anxious and defensive individual (Beasley 2008a; Burkitt 2012; Holmes 2010).

There is considerably more room for manoeuvre, at least with regard to embodied emotion, in so-called 'pro-sex' feminist literature which specifically resists a calculating conception of the self and singularly risk-oriented approach to sexuality/sexual relationships, instead opening up space for uncertainty, adventure and even fun (Beasley 2013; Beasley et al. 2012).[2] While this scholarship precisely turns on a rejection of narrowly cognitive and risk-focussed agendas for women in particular, it nevertheless tends to emphasise forms of individualism and libertarianism that are not especially conducive to understanding *social* subjectivities, interconnection, innovation and change (Bracewell 2016; Glick 2000; Sorooshyari 2011). Since sociality is central to our discussion of internet dating, we turn to the concept of emotional reflexivity to forward our thinking.

In using the term emotional reflexivity we differ here from some major approaches to reflexivity such as those promulgated by Anthony Giddens and Margaret Archer (Archer 2007, 2003; Giddens 1992, 1990). In our reading of the terminology, emotions are not the interior property of individuals but rather involve a mode or aspect of sociality, a sociality shaped by competing discourses of power which constitute subjects. The competing character of discourses, which are embedded with and composed of emotionalities, offer the means to a variety of forms of subjective investment and conduct, enabling the creation of what Delueze and Guattari describe as deterritorialising lines of flight from the horizon of hegemonic discourses—for example, in relation to gendered heteronormativity. Emotional reflexivity thus refers to the process via which subjects may undertake an 'undoing' or unsettling of emotionalised discourses. This undoing may be enacted through relational intersubjectivity and involves both idiosyncratic emotional interpretation and practices which can generate potential departures from dominant discourses and upset the notion of fixed feeling rules. In this context, rather than arising from some quality of individual self-agency intrinsically prior to social context, social change emerges from heterogeneity and inconsistency between discourses which interrupts the reiteration of hegemonic articulations. In this context, emotions are profoundly social practices. Instead of being a feature of individual interiority, emotions are played out in the intersubjective relationality of the social world. In this analysis we suggest not only a rather different view than that of Giddens and Archer but, in drawing upon Foucauldian accounts of power as productive including productive of subjects and accounts of social change outlined by Deleuze and Guattari, we also explicitly highlight (perhaps even interject) emotions into their formulations (Foucault 1980, 1977; Deleuze and Guattari 2004, 1987). In our understanding of emotional reflexivity, power and discourses are perforce composed of and through embodied relational emotionalities—rather than emotions merely being produced by them. Power and discourses have emotional substance shaping how subjects can navigate and enact modes of (socially recognisable) feeling (Beasley 2012; Holmes 2010).

However, in the context of scholarship on internet dating which commonly concentrates on defensive emotional responses, the place of excitement in emotional reflexivity deserves further examination given its connections with positive emotional engagement and generating directions in progressive social change. Excitement is here defined as a spontaneous and elementary feeling 'inimical to the orderliness of life' (Elias and Dunning 1986: 72). Elias and Dunning (1986) argue that this playful form of excitement becomes increasingly important as a form of compensation for the generally tight control of emotional expression expected within complex societies. This account is limited by its reliance on a repressive model of power, a model challenged by Foucault's contention that power produces and not simply represses action (1990). Nevertheless, Elias and Dunning (1986) provide a useful account of excitement as subject to increasing control and regulation which is not necessarily to be regarded as at odds with Foucauldian insights. As discursive boundaries and social rules shift in terms of where, when and how excitement is 'allowed', greater emotional reflexivity is required in interpreting one's own and other people's emotions and using them to feel one's way through the world.

While the approach taken by Elias and Dunning proposes that emotionality including excitement is increasingly restrained in modernity, this approach does not distinguish between aspects of emotionality. There is little allowance here for possible signs of increasing discursive contestation over emotionalities and in this context growing forms of emotionality in certain settings, for example, between men and their children in some modern contexts like Sweden (Hearn et al. 2012; Klinth and Johannson 2008), and indeed between men and women in the enhanced emotionality of developing intimate relationships that arises in internet dating. Moreover, we would suggest that a more nuanced perspective which pays attention to diverse aspects of emotionality, and their uneven development over time, can bring into clearer view the impact of increasingly less fixed social norms and expanding tensions between inconsistent or conflicting normative discourses regulating emotional conduct. Stronger emotional connections with others in the face of normative social uncertainties do not of themselves generate social change in terms of remaking social hierarchies or producing more egalitarian relations, and specifically do not inevitably advance gender egalitarianism in heterosexual partnerships. It is not uncommon, after all, in histories of emotion to feel strongly about those one deems to be one's inferiors or dependents (Beasley 2017b). By comparison, a growing requirement to deal with discordent discourses and to *interpret* contexts of emotional sociality in the face of less clear-cut, less generalised and less consistent social norms of interconnection has *the potential* to reorder elements of heterosexual relationships towards greater gender equality (Holmes 2015b). In this setting, the heightened emotional experience of excitement in internet dating becomes not so much about having to 'compensate' for having in general to discipline one's emotions in modernity—a view which tends to presume some a priori conception of an intrinsic human amount of emotionality—but rather possibly offers a means to navigate competing emotional discourses as they arise through relating to others and to deploy reflexivity in enacting changes in ways of relating.

If the undifferentiated account of emotions put forward by Elias and Dunning is not taken as given, it then becomes possible to recognise that, for example, emotional involvement/connection/bonds and emotional reflexivity are by no means identical processes and may well have different implications for social change. Yet, they may also augment each other. For instance, although stronger emotional connections do not necessarily advance social egalitarianism, it is possible that the labile setting of enhanced emotionality associated with embarking upon intimate relationships may well provide more motivation for greater emotional reflexivity. In this context, the engagement of intense emotions like excitement regarding possible intimate partners may catalyse reflexive reconsideration of social norms This may occur at the level of individuals' intimate relationships and practices, but *en masse* it can form the basis for encouraging micro-political shifts in how, for example, people deal with institutions and other social assemblages but also how they relate to others. In this case our interest is in gendered heterosexual relations and what role excitement might have in how they play out in internet dating. To investigate that role we need first to question dominant representations of internet dating as a form of rational choice that seeks to find the best product, in what is often described as a marriage or mating market.

Shopping for love? Beyond risky markets to exploratory fun

There are ways in which internet dating brings intimate life into the commercial realm, but it is debateable to what extent this determines how people experience it. There is certainly no doubt that internet dating is a profitable business. In the UK in 2020 revenue for internet dating amounts to £119 million and around $64.4 AU million in Australia (Statista 2020a). However, it is not possible to discover a causal relationship between the money made by companies offering services and the feelings and practices of users. Long before internet dating, previous forms of dating have been commercialised to some degree by the marketing of various romantic experiences such as dinner or movies or Valentine's Day (Heino et al. 2010: 431). Moreover, commercialisation of services matching couples is not confined to America or to recent history (Barraket and Henry-Waring 2008; Hardey 2002; Whitty et al. 2008). There are those who have argued that consumerism has made relationships more risky in late modernity. Some speak of 'throwaway relationships' in a wasteful society (Bauman 2003), or 'pure relationships' in which parties negotiate intimacy and exit as soon as it is no longer felt to profit them (Giddens 1992). Others less histrionically analyse the triumph of a 'spirit of consumerism' over a previous Romantic ethic more interested in passion than pennies (Campbell 1987), or give voice to this supposed triumph by simply accepting competitive 'rational choice' (often understood as fuelled by biological imperatives) as *the* central way to understand internet dating (see Bruch and Newman 2018). These accounts have been subjected to robust criticism, not least by leading feminist scholars of intimate life who have pointed to empirical evidence suggesting that relationships often still operate according to rather more feudal, less negotiable rules that entrench inequalities between women and men, children and parents (Jamieson 1999; 2013; Smart and Shipman 2004). However, these feminist criticisms still seem to focus on the continued riskiness of relationships, including those started online.

Some feminists specifically consider the commercialisation of intimate life and how it might render relationships subject to new risks. Hochschild (2003) has argued that relationships in the private sphere have come to be governed more and more by rationalising principles of profit, efficiency and calculability. She suggests that the kind of rationalisation that has occurred in the workplace is entering the private sphere of relationships. Thus, as at work, rational and profit principles are allegedly spilling over into intimate lives. This would mean that people are required to navigate this newly dominant discourse by managing themselves and their feelings in intimate relationships in similar ways to paid workers with presumably similar problems of having to surface or deep act to conform to such feeling rules. In this analysis the 'marriage market' increasingly parallels the job market. Hochschild (2003: 13–29) has explicitly suggested that for women, the new norms around relationships advise a 'cooling' in how they do intimacy. Less investment in the emotional needs of others is prescribed by dominant cultural discourses, Hochschild asserts. She criticises the alienating consequences of this masculinist model and implicitly mourns the stepping back from previous caring norms associated with women's role in relationships. As always, there are questions about to what extent practices follow prescription (Seidman 1991: 6), but for Hochschild the shift towards marketising sex and romance is clear. Illouz (2013; 2006) is similarly pessimistic about the impact of commodification, viewing this as constituting part of a process of devaluing of women that has emerged as a result of the sexual revolution lowering the cost of sexual access for heterosexual men. She sees internet dating as a prime instance of marketising sexuality and love in that it promotes asessments of worth based on visual, fast and asymmetrical evaluations of women's sexualised bodies. This marketising relies on youthful attractiveness. The supermarket-like abundance in internet dating is argued to diminish the value of women's bodies and to create pools of women considered 'unworthy' by these standards (see also Heino et al 2010; Hochschild 2003).

However, accounts of the practices involved in internet dating vary. Not all feminist or other commentaries conceive of it as irretrievably immersed in the marketisation of intimacy (Miller et al. 2016: 21). Additionally, it is important to note that accounts which treat all dating sites and their associated algorithms as much the same ignore considerable variability in terms of emphasis upon physical appearance, often linked to the age profile of users on particular sites. Such overarching accounts ignore the specific development of some sites and site features which work against simple visual assessment. While apps like *Tinder* and *Grindr* largely match daters on the relatively reductive basis of geographical access and physical appeal, many other sites do not have this orientation. Thus critical attention to *Tinder*, for example, must be viewed as a particular focus rather than as revealing something intrinsic to all online dating.

Tinder has the reputation of being a hook-up app, is primarily for younger users and, in the USA, UK, France, India and Italy at least, is dominated by young men. Indeed, twice as many American men as women employ this app. Iqbal's assessment of data on *Tinder* daters indicates that young men users are less discriminating about who they contact, quicker to send messages, less verbally communicative, far more interested in one-night stands and overall more instrumental in terms of pursuing physical attraction, by comparison with women using the site (Iqbal 2020; see also Gatter and Hodkinson 2016). *Tinder*'s reliance on photographs as almost the sole basis for dating selection decisions is linked to an objectifying focus on women's appearance but also, as Pascoe notes, the control over self-presentation afforded by its format helps young men to manage the 'profoundly unmasculine display of vulnerability' that is frequently a feature of 'flirting and getting to know someone' (2011:17). Similarly, *Tinder*'s format is inclined to facilitate homosocial bonds and the performance of masculinity as this participant in a qualitative study outlines: '[i]t's quite common for guys in relationships to be crowded around the single guy and his Tinder account and having fun. Helping him to write messages which are very . . . not what they would normally send' (*Francis, UK*—cited in Haywood 2018: 141).

While marketised shopping for dates may be a way to describe young men's use of the app, we suggest that it is at least as helpful to draw upon scholarship regarding gender and masculinity to explore *Tinder*'s fit with the gender stereotypic practices of its users (Hess and Flores 2018; MacLeod and McArthur 2019; Thompson 2018; Tiffany 2020). The *Tinder* format, and the practices it supports and encourages, do not appear to advance gender innovations (Lee 2019). However, even if *Tinder* might be deemed an archetypal site to discuss marketisation of intimacy, it should also be acknowledged that this form of dating site is used for other purposes than hook-ups, such as developing self-affirmation, longer-term relationships and friendship. Practices of intimacy may be heterogeneous, no matter how particular technologies of intimacy are constituted (Aretz et al. 2010; Ruggiero 2000; Stafford et al. 2004).

Whatever the degree to which *Tinder* can be viewed as illustrating increasing marketisation, it is crucial to keep in mind that *Tinder* is different from many other sites. For example, '95% of *Tinder* users meet their matches within a week (compared to 25% of online daters overall and 15% of offline daters)' (Iqbal 2020). The rise of sites, like *Inner Circle* and *Hinge* which specifically reject the *Tinder* swipe feature and aim at developing connection, along with sites that have intentionally added features to encourage more substantive interactions (such as *OKCupid*), provide evidence of a shift away from *Tinder*'s rapid hook-up format and of online daters becoming 'Tinder tired' (Dixon 2020).

Moreover, although use of market metaphors by users is common (Heino et al. 2010; Whitty 2008c), this does not mean that the rules of the commercial sphere become dominant in guiding practices of intimacy online. Discourses promoting 'true' love as something found serendipitously, compete with exchange-oriented models in which 'people are reduced to deuhumanizing lists of attributes' (Baym 2015: 97). The work involved in managing profiles and presentations of self has already been mentioned (see Chapter 2) and this may sometimes be thought of in terms of selling oneself (Jagger 2001, 2005; Whitty 2008c). The fact that users are paying for access to other's profiles would also seem to encourage people to think about their internet dating in relation to money and markets. Internet daters do describe sites as like a supermarket or catalog but this may be said in ways that are intended to be amusing and often highlight 'shopping' as a pleasurable activity. For instance, men may describe it as like being 'a kid in a candy store' (Bean 2011: 189). And some women may talk about seeking men who 'tick all their boxes' *(Lucy, UK, dating blog, 2019)*. As this mention of ticking boxes indicates, however, not all list-making should be presumed to inevitably draw upon a commercial view of social interactions. Advice in an Australian dinner dating site newsletter insists that developing lists of one's interests and values is crucial to understanding who you are and who might be a good match. Here, lists are deemed to be an aspect of taking responsibility and of personal agency, rather than a self-serving means to procure what you want (Margaret 2020). Apart from a possible range of meanings attached to list-making which disallow any immediate assumptions about commodification of date selection, many daters also actively resist a calculating shopping approach to meeting potential partners, as we will shortly outline in more detail.

Unlike interview-based studies (Heino et al. 2010; Whitty 2008b: 1714) , most of the internet daters in our sample of online accounts do not describe their practices in language akin to shopping or selling. It may be that in interviews, '*when asked* they presented a shopping list of what they were seeking' (Whitty 2008b: 1716, our emphasis), or used shopping metaphors (Heino et al. 2010), but in the self-organised stories about experiences of internet dating we sampled there are relatively few mentions of a 'shopping list' of requirements:

> [o]ne woman on OKCupid, who was married, sent me an email with a 5 point list of things she wanted to do to me; normally, at this point I'd gloss over exactly what was on her depravity shopping list, but . . . (LAST CHANCE TO ESCAPE TO KITTEN VIDEO BEFORE SLEAZE AND HORROR) . . . since I've given you enough chances to look away, it can basically be added up to her husband pissing on me while she choked me, while I was locked in a cage in their sex dungeon.
>
> *(Jacob, UK, individual blog, 2013)*

Here, the 'depravity shopping list' is humorously related, rather than the writer seeming to seriously be morally outraged. Yet, this might reveal the affordance of internet dating in providing the opportunity to ask for forms of hetero-sex likely to be seen by most as beyond the normative. Another dater shares with the blogger Lucy, again with a mixture of horror and humour, a story about a woman sending him an eight-item list of what she likes with 'anal' and 'kink' as the first two items and 'my daughter' as the eighth. As Lucy's date comments: '[i]f this is in order of preference she clearly likes anal much more than she likes her daughter . . .' *(Lucy, UK, individual blog, 2019)*. This comment implies that those who do seek online dates by employing very specific 'shopping lists' in such rapacious ways may be 'already distanced from conventional moral economies' (Ross 2005: 347–349). Moreover, as mentioned earlier in relation to *Tinder*, those who adopt highly instrumental modes of internet dating which reduce meeting potential dates to 'shopping' are likely to choose particular kinds of dating sites and/or to adopt such an approach at a particular time in their lives. A narrowly commodified approach to dating did not appear to be the main orientation, let alone the sole or intrinsic one, of internet dating site users in our research.

While recourse to the shopping/buying metaphor may be found in daters' accounts, as we have intimated there is also resistance to this trope and to doing internet dating in calculating, 'profit'-maximising ways (Heino et al. 2010: 440–441). For instance, one blogger is rather dismissive of *Ashley Madison*'s 'fairly exhaustive list of sexual checkboxes' precisely on the basis that it resembles a shopping list *(Jacob, UK, individual blog, 2013)*. Another person commenting in response to an online article relating an internet dating 'success story' says, '[m]y personal pet hate is men who list their attributes as if they were a car. You know the type: only one previous owner, low mileage, in great condition. Aggghh!!!' *(Commenter on Phillip's story, Australia, online newspaper story, 2011)*.

The writer, who we might presume is a woman, disparages men who write their profile of themselves as if they were advertising a product. There is no elaboration, so there may be a number of reasons why this irritates the commenter, from the rather clichéd association it draws on between masculinity and cars, to the way in which it directly uses objectifying sales language. Her dislike of this practice seems related to the way it implies she is shopping for a man in the way she might shop for a car. Caution in applying market metaphors is also evident in the comments of a writer discussing profiles. Phillip admits he 'treated [his] dating profile like a bit of a resume' *(Phillip, Australia, online newspaper story, 2011)*. This does conjure up an image of the internet dating site as akin to a job market, but it is worth noting that this is a passing comment and it is mentioned rather apologetically. There is little here to indicate casual acceptance of marketisation. Similarly, a friend of Jacob's takes issue with a rather literal version of the 'meat market' metaphor (Heino et al. 2010: 428) and in so doing offers a humourous appraisal of it. Jacob reports her account of eventually meeting a partner online via *Muddy Matches*, an agricultural/country dating website.

> There were some astounding men on there, and not in a good way. Having been assessed by some as to whether I'd breed well (estimated breed value ebv is a common term in ag). I was lucky to find [my partner]. If you're in need of entertainment then it's worth having a look!
>
> *(Phyllis, cited by Jacob, UK, individual blog, 2013)*

This woman's report is dismissive and critical of the commercial valuing of her evident in being explicitly 'assessed' by some men using agricultural breeding terms usually used to evaluate livestock. Although slightly different from the usual sexualisation of women as pretty objects, this kind of estimation of the 'value' of women when searching for partners is akin to other forms of shopping. Such ways of thinking are not new to the internet, even if the internet is more likely to be viewed as an online superstore than a local country fair. However, the suggestion is that rather than being a target for feminist outrage, the woman regards this site as entertaining because of the over-the-top emphasis on breeding. There is pleasure to be had from laughing at the quirks and foibles of others as they present themselves to other internet daters.

If internet dating bears a resemblance to consumption perhaps it is not always or only about making calculated lists of desirable qualities in a partner, but more like the many ineffable and social pleasures of shopping (Falk and Campbell 1997). This may include, as suggested above, the amusement drawn from how others act or the excitement of the 'endless possibilities' (see Chapter 3). Internet dating may *sometimes* be fun (Zytko et al., 2018).

> I've used it for almost 3 years (on/off) in addition to the bar scene, etc. & had an absolute fun in doing it (sic) . . . it IS FUN when you meet new people, even if your purpose is different to others :)
>
> *(Commenter on Phillip's story, Australia, online newspaper story, 2011)*

Meeting other people may well be pleasurable, even if the end goal of finding a partner is not achieved (see also Chapters 5 and 6). Many of the accounts mention enjoying this aspect of internet dating and advise others to do likewise, as is mentioned by another commenter on this same newspaper story: 'Remember to have fun. If the date doesn't work out, or they are a freak, you can always have a laugh about it later' *(Commenter on Phillip's story, Australia, online newspaper story, 2011)*.

Here, the enjoyment is not just about meeting other people, but also being able to laugh about the dates that go badly or the people met who turn out to be 'a freak' (see Chapter 2). However, what is already evident in the quotes given is not only resistance to commodifying self and others but some care for others, for example warning them to look away before discussing sexual 'depravity'. The humorous tone is a reminder that these accounts are clearly meant to entertain, but they also convey stories of excitement.

Internet dating is exciting (sometimes)

These accounts of internet dating do speak of it being exciting, especially looking forward to meeting up, the actual meeting and the frisson of physical proximity and contact. For example, one woman admits she was 'actually quite excited' when she arranged to meet a date the next day *(Sheila, Australia, dating blog, 2014)*. Similarly, Peter describes going to Moscow on his first internet-facilitated date. They took a taxi at the end of their first evening:

> her hand touched mine and. . . . She leant over and whispered in my ear. . . . Now, you may be thinking 'so what'? You have to remember at this point in time that I hadn't had any degree of intimacy with a woman for months, if not years, so that actually made me quite excited.
>
> *(Peter, UK, individual blog, 2005)*

For this man, even a small touch and a whisper in his ear is thrilling. This may make us consider again how heterosexual pleasure and arousal might often take non-penetrative forms that are nevertheless important to people (Beasley et al. 2012: 89). However, while these forms of excitement often begin in online exchanges and during the stimulating uncertainty of the interval prior to the first meeting, they typically reach their zenith in the delicious promise offered by offline face-to-face meetings.The online stage of internet dating is more specifically marked by the cycle of excitement and disappointment mentioned by some of the accounts in our sample. Searching for a partner online is not simply a rational decision to 'shop' for a partner in a way that 'increases temporal efficiency' (DeMasi 2011: 211), it is an emotional experience. Excitement comes with the almost endless possibilities of finding new partners (casual or long-term), and this may be fuelled by people who meet online initially idealising potential partners (Whitty 2008b: 1708) such that disappointment often results from the face-to-face dates. Yet, sometimes disappointment may come as part of the online stage. One young woman describes her engagement with internet dating as follows:

> I would call it 'love-hate' type of thing: you go into it - get super excited - enjoy it for a while - get disappointed - delete your profile - live without it for a while - do your day to day - get bored - go back to it - and here we go again!
>
> *(Annie, UK, dating site blog, 2015)*

In this cycle the excitement comes at the beginning when 'you go into it' and the disappointment sets in after you have enjoyed it for a while. All the trawling through profiles and messaging that leads nowhere can contribute to 'the mind-numbing tedium and eternal disappointment of dating apps' *(Lucy, UK, individual blog, 2019)*. Yet, boredom with your everyday life and the hope of finding 'the one' send these daters back looking for more. As Andy, a twice divorced man living in London who had been using *Tinder* for some years explains, the hopeful excitement at the beginning is intoxicating, even if the fallout from excitement is deflating and indeed deadening.

> The fallout is the trip home, where you feel crushed and defeated and you know you're going back to square one to start . . . again.
>
> [I]t's very soul-destroying. The addictive part tends to come mainly from . . . trying to find someone to keep us company.
>
> *(Andy, UK, online news story, 2016)*

Many daters feel excited by the possibilities for romance when they begin to date, by getting emails from people they are attracted to and by the thrill of online flirting (Frohlick and Migliardi 2011: 43), but the excitement can fade as dating becomes routine, rather than retaining the special aura expected of romance and sex (Jackson and Scott 2004). This process is described by a male blogger who appears to be in his thirties at the time.

> Having e-mailed each other extensively and entertainingly for about a month we agreed to meet up for drinks. I have to say that I wasn't really very excited beforehand, but that was probably due to the fact that by this point I'd been on so many dates in a relatively short space of time.
>
> *(Steven, UK, individual blog, 2011)*

This is a young man who dates so frequently that the sense of excitement he may previously have had before meeting a woman has disappeared. The volume of dates facilitated by internet dating, for him has rendered dating rather routine and unexciting.

Similarly, internet dating does not always excite risky movement beyond heteronormativity and may be boring and frustrating for women in ways that reproduce traditional gender relations. One young woman blogger says she can't 'get past some of the lingering gender-based rules' and that as a woman she 'plays a passive role, the receiver of attention, the awaiter of messages' (*Elizabeth, Australia, lifestyle blog, 2013*; see also Smaill 2004: 96).

When internet dating is exciting, it is not inevitably transformative of relations between men and women, although it may encourage emotional reflexivity. For example, little change results from the internet dating cycle of excitement and disappointment where the excitement of endless possibilities for romance is tempered by the reality of many boring dates with men and women—who are often constrained by gender norms and prejudices. Yet, these accounts do hint that emotional reflexivity is promoted because many of these internet daters describe ways in which their experiences have made them re-examine their feelings, their actions and themselves. This is perhaps the most explicit statement of such processes.

> Most importantly, online dating taught me lots of things and most of them about myself! The knowledge you get about yourself in the process is priceless. I found out what my limits and limitations are, what I like and hate in other people's behaviours, how I come across to others (thanks to the feedback of some nice people), and how I react in different tricky situations.
>
> *(Annie, UK, dating site blog, 2015)*

The likelihood of multiple dating experiences, which typically arise in undertaking internet dating, furthers possibilities for such reflexivity as several parallel or chronologically close experiences arise to provide opportunities for comparison. Different and often competing emotionalised discourses regarding dating, romance, desire, love and coupledom afford and even require navigation. Thus, Lucy *(UK, individual blog, 2017)* is worried when she is about to meet an 'online dating virgin' whether he will appreciate that '[i]t's not going to get better than this for him', without other bad internet dating experiences to which he can compare it. The opportunity for immediate comparison is less likely with offline dating, where serial monogamy remains a dominant discursive norm, even amongst the young (Regnerus and Uecker 2011). Thus there are some indications that internet dating might prompt reflexivity and provide some space for innovations that go against the gendered heteronorm.

Internet dating excites non-normative directions

Internet dating can offer excitement through divergence from predominant norms of seeking and meeting familiar others, as 'it gives people options to find people in other social circles that you may not ever cross paths with' *(Commentor on Phillip's story, Australia, online newspaper story, 2011)*. This access to strangers increases opportunities to encounter less familiar discourses and to undertake somewhat less conventional practices. We deliberately refer to *divergence* here because, rather than strongly challenging, let alone upending usual homophilic/homogamous expectations that people will find partners from within their communities and personal networks, internet dating appears to provide opportunities to gently shift such boundaries. Homophily/homogamy are terms which describe people connecting and partnering respectively within the social groups to which they already belong. The terms overlap within the field of internet dating since connecting/bonding (homophily) and developing partnerships, including marriage (homogamy), with those who are like you are not discrete processes.[3] However, for simplicity we will only refer to homogamy from now on as this term is more specific and more clearly focusses upon the generation of intimate partnerships. Given internet dating's imperative to develop such partnerships (even if other social connections may be instigated), this terminology allows us to consider how this relatively new technology of intimacy might possibly enable divergence from homogamy.

The internet allows people to seek out, meet and develop intimate partnerships 'with perfect strangers, that is, people with whom they had no previous social tie' (Rosenfeld and Thomas 2012) and at the same time more and more people are seeking partners through internet dating. Online dating is claimed by some to have shifted the overwhelming pattern of homogamy across the globe (Hergovich and Ortega 2018: 1–30; Rosenfeld and Thomas 2012). However, the degree of this shift has also been questioned. In the first instance, the algorithms employed by many internet dating sites match daters precisely on the basis of similarity (Finkel et al. 2012: 20–21). Second, while homogamy may be reduced by internet dating, most people continue to contact and meet people who are fairly similar to themselves in terms of race/ethnicity, level of education and religion (Hitsch et al. 2010; Ciabattari 2017: 95). Ciabattari further suggests that, even though there are some features of internet dating that could assist in reducing homogamy, this shift may well be uneven. For example, she notes that some groups, which have more restricted pools for intimate partnership, do appear to gain from the expanded options available through internet dating which extend far beyond their usual social circles. She, among others, points to daters attracted to same-sex partners and middle-aged heterosexuals as clearly benefitting from the opportunity not be restricted to dating partners from similar backgrounds (Ciabattari 2017: 94–95; Hillier and Harrison 2007; Rosenfeld and Thomas 2012). It is worth remembering here that for lesbians and gay and bisexual men, meeting sexual and romantic partners using online methods has become prevalent, even ubiquitous, in countries like Australia, the Nordic countries, China, the USA and many others (Anderson et al. 2020; Clue and Kinsey Institute 2017; Cummings 2020; Geller 2017; Miller 2019; Prestage et al. 2015). Internet dating is evidently a successful means to overcoming the limits of social homogamy for groups whose familiar local circles do not provide a wide range of dating options.

The question is whether the capacity of internet dating usage to enable divergence from homogamous norms goes beyond those restricted to thin dating pools. Since online daters tend to meet complete strangers, the growing use of internet dating worldwide has the potential to encourage social connections between diverse groups in modern societies and globally, perhaps thereby ameliorating social divisions. This potential is a matter of recent debate and not by any means straightforward. Yet, even if internet dating's role in reducing homogamy is not entirely clear in the literature on intimate partnering, that literature does regularly associate homogamy with the maintenance or deepening of social divisions (Brynin et al. 2008). Hence, if online dating does have any appreciable effect upon norms of homogamy for mainstream daters as well as specific minority groups, this has significant social implications for this book's guiding question regarding whether and to what extent internet dating might advance social equality. In this setting it becomes relevant to consider partnership choices in our sample of internet daters in the light of scholarship on homogamy and online dating.

Homogamy is frequently seen as a marker of and enabling the reproduction of social inequality (Furtado 2015). The classic link between homogamy and inequality arises in relation to race/ethnicity (Feliciano et al. 2009; Robnett and Feliciano 2011; Yancey 2009, 2007). Researchers discussing the USA, for example, note the low number of interracial/interethnic marriages in that country, while drawing attention to increases in such marriages in recent times. In this context, Hu and Nash (2019: 312–313) draw attention to a significant *decline* in spousal racial homogamy within the United States over time from the early twentieth century onwards (see also Lamidi et al. 2015; Wu 2018). Relatedly, there has been a striking drop since 1990 in the percentage of non-black adults in the USA opposed to these marriages (Bialik 2017; Livingstone and Brown 2017). It has been suggested that, given the growing number of marriages that begin online, this is likely to increase interracial/interethnic marriages and thus assist in reducing racial/ethnic divisions which contribute to social inequality. Hergovich and Ortega point out that in the USA, following the development of dating websites such as *Match.com*, during the 2000s interracial marriages increased from 10.68 to 17.24 per cent in 2015 (Hergovich and Ortega 2018: 3–4). Even if a direct causative link between internet dating and this decline in US racial homogamy may be difficult to confirm, it does not seem unreasonable to allow that internet dating could contribute to this rise in interracial relationships and thereby assist in mitigating social divisions and inequality associated with racism.

Nevertheless, since research on racial/ethnic patterns in partner selection has been predominantly studied in the United States, comparatively little is known about whether these patterns are also found in other countries with different demographies and historical traditions (Potârcă 2014: 11, 62). There are, for example, marked variations in these patterns even between European countries and within different European communities (Potârcă and Mills 2015). Notwithstanding the continuing prevalence of racial/ethnic homogamy in online partner selection across Europe, Potârcă argues that 'as opposed to the American context' there is 'a greater inclination towards assimilation among minorities' in European countries (2014: 62). While her study does not include the UK, this research suggests that racial/ethnic patterns in partner selection may be different and perhaps less clearcut in the UK than in the USA. UK partnering trends do not mimic those in the United States. They arise from markedly different populations and legal/cultural traditions (Mohdin 2018). In this setting, a multitude of UK print and online sources have insisted in recent times that interracial relationships are now widely accepted in Britain following the engagement and then marriage of Prince Harry and Meghan Markle (Caballero 2019). Such public narratives are undoubtedly over-confident (Moss 2020; White 2019) . However, they may indicate shifting patterns of intimacy (Bingham 2014). Lampard states that online settings in the United Kingdom are generally less associated with homogamy than equivalent offline contexts (2020: 17), while acknowledging that some offline contexts are also compositionally diverse. The analysis suggests a somewhat variable relationship between homogamy and online connections and points to the difficulty of establishing causal links, yet notes an overall tendency towards racial/ethnic heterogamy in online partnering (Lampard 2020: 8).

In a multicultural country like Australia, partnering outside one's cultural grouping is seemingly not as fraught or as radical a path as it is in the USA. The 2006 Australian Bureau of Statistics data on inter-partnering (interethnic/interracial marriage) following migration indicates that by the third generation this becomes a 'common occurrence' (SBS 2013a). By 2016, 31.6 per cent of marriages in Australia were between people from different countries of birth (Watts 2019: 104). Khoo et al. (2009) point to a marked increase in inter-partnering across generations in all ancestry groups, with a significant jump in inter-partnering amongst those of Greek, Lebanese and Chinese ancestry. On Australia's largest internet dating site, *RSVP*, two thirds of users in 2012 did not rate ethnic background as important when identifying a suitable partner (SBS 2013a).[4] Thus, while the three national contexts of the United States, United Kingdom and Australia may show differences regarding a decline in homogomous partnering, both the US and Australia show a significant rise in heterogamy, with the UK indicating some evidence of this rise (Potter-Collins 2014; Song 2015)—a rise taking place during the concomitant increasing use of internet dating. Since internet dating would appear to advance intimate and other connections between those who did not previously know each other, the massive growth in use of online dating sites appears to be of potential benefit in undermining forms of homogamy which support social division and inequality.[5]

It is important to note here, however, that the term 'homogamy' is not a synonym for inequality. Indeed, in some instances, intimate partnerships between people who are similar to each other and who belong to similar groups may actually be a means to advance social equality. For example, in marginalised groups partnering *within* the group can maintain endangered cultural/religious or other bonds and resist assimilation into mainstream culture. Second, increasing age parity in intimate relationships between women and men (for example, reductions in significant age disparities in intimate partnerships between younger women and older men), is likely to support gender equity (see also Chapter 6). These two instances illustrate the ways in which homogamy (similarity in partner characteristics) can sometimes be associated with advancing equality and can work against disempowerment of some minorities (such as Indigenous and Jewish groupings) and women (Hu and Nash 2019; SBS 2013a). Nevertheless, homogamy is much more usually associated with upholding and extending social inequality, as a crucial key in the reproduction of privilege and disadvantage (Blackwell and Lichter 2004; Waldi and Byron 2006). The problematic impact of homogamy is, for example, typically evident in homogamous partnerings in which people from similar racial/ethnic, class, educational and religious backgrounds develop intimate relationships. Partnerships between those who are alike in these characteristics shore up socio-economic and cultural capital. Thus, if intimate partnering in internet dating diverges from homogamy, it can help challenge sedimented forms of inequality.

Online relationships of all kinds might be more likely to cross social categories, as there is some evidence that cross-sex friendships, interracial and age-dissimilar relationships are all more common online (Baym 2015: 99). While, as noted earlier, there is a considerable body of research indicating the particular importance of being online in extending the pool of potential same-sex partners, especially for young people in small or conservative communities (Hillier and Harrison 2007), there remains nonetheless a need for better understanding of the benefits and excitement of such heterogamy to heterosexual users. For example, although age disparities between younger women and older men can be viewed as at odds with homogamy, in this case heterogamy is normatively gendered and unlikely to enhance heterosexual egalitarianism. By contrast, age disparities in which older women pair with younger men involve some divergence from heterosexual norms and may shift existing gendered power relations to the benefit of women. Older women might use internet dating to avoid pairing up with older widowers they know, whom they might end up having to care for in conventionally gendered ways (McWilliams and Barrett 2012: 426; Frohlick and Migliardi 2011: 83).

The point here is that internet dating can provide a means for heterosexual daters to diverge from the many forms of homogamy which are associated with inequality (such as socio-economic, religious, lifestyle, status, cultural, professional, educational and certain modes of racial/ethnic homogamy), as well as sometimes diverging from particular heterogamous norms which support inequality (such as norms concerning older men partnering much younger women). In both instances it would seem that online dating has social equity potential.

Internet dating certainly offers relatively easily available opportunities for users to find partners who are different from themselves in a variety of ways. One of the accounts in our sample recommends this, although it is worth noting that the potential to connect with the unfamiliar is not straightforward and may involve fear: '[g]oing on a date with someone different scared me and put me out of my comfort zone, but now that I've done it once I would definitely recommend stepping away from your usual type and giving it a go' *(Orla, UK, dating site blog, 2015)*.

She describes the fear related to 'stepping away from your usual type', but the fact that she recommends others do the same suggests that being scared of this risk of dating someone different was by no means an entirely negative experience and indeed had elements of excitement to it. She notes her previous homogamous behaviour in saying that in selecting partners in the past she has chosen 'someone who is suited and booted, typically working in a similar type of job'. Her divergence from this pattern has made her think that it has been 'interesting to hear about a different side of life' and she says that her new partner was 'worth the leap of faith'. It is not inevitable that she will eschew her 'corporate' life for his more 'alternative' one, or that such individual pairings will bring wider social change around class and occupational hierarchies. However, if internet dating facilitates more diverse pairings on a larger scale there is some space for it to trouble or undo some of these hierarchies. It is interesting that this example is not the perhaps more familiar one of women engaging in hypergamy (partnering 'up' the social hierarchy) as a route to individual social mobility (Stacey 2004: 185). Here, the woman is the one with the corporate career and the man she meets has more seasonal, less prestigious work and is 'a great cook'. In other cases, it may prompt people to find partners from different ethnic groups, as it does for the man who 'hadn't considered dating someone that was Asian before', until he met his Chinese partner of two years online (*Bruce, Australia, online dating story compilation, 2014*).

Although as noted earlier, internet dating research generally suggests that most online daters continue to look for partners of the same race/ethnicity and education level (Rudder 2014a, Whyte and Torgler 2017), there are nevertheless excitements attached to internet dating that have unconventional possibilities. One of these is the way it enables transgression, or even a limited subversion, of monogamy. First of all there are some technical features of online dating which enable users to be directly aware of the potentially unsettling, possibly exciting, but also ubiquitious practice of parallel dating.

> Most apps put a time stamp on everyone's profile, so that you can see when anyone has last been logged in. For example, you could find out if the man you went on a date with last night was looking for other women while you popped to the loo in the middle of dinner (he was).
>
> (*Anonymous dater, UK, newspaper story,* 2015)

Moreover, as we mentioned in Chapter 3, norms of monogamy in online dating are less settled at least during the equivocal period before partners clarify that their relationship is serious and exclusive, usually accompanied by the decision to take down dating profiles and leave the dating site. Monogamy during this equivocal period is also challenged by a peculiarity of internet dating—that is, the emergence of forms of what we have termed 'mundane polyamory' (Beasley et al. 2018; Beasley and Holmes 2016). This challenge to heterosexual orthodoxy is not necessarily to be equated with 'cheating'—the usual understanding of non-exclusive behaviours in off-line dating—in that internet daters are well aware of the commonality of at least a short period in which daters increasingly become more involved romantically and sexually with a particular individual, but may still maintain a profile and thus continue to consider, message and meet others as a part of the usual online dating experience.[6]

Online 'dating' of a number of people at once is an example of transgression from established offline dating practices. Transgression denotes practices that wander temporarily, and not usually very consciously, from the heteronorm (Beasley et al. 2012). Such transgressive practices sometimes become evident in a high frequency of dates. Interestingly, in these instances transgression can make dating feel less special and therefore less exciting, as described earlier. It is thus open to question whether meeting with many possible partners (casual or long-term) is always thrilling. The mundane aspect of it is evident from the often very casual way in which it is mentioned in online accounts. For example, a long-term internet dater talks about going on a date and spending the night with her, and then mentions without any fanfare that they 'were waiting on the train platform the next morning as she was going home and I was actually going to meet Date 83' *(Steven, UK, individual blog, 2013)*.

Even the thrill of more familiar forms of non-monogamy is in doubt. There has been some comment on extramarital affairs, and the appeal and morality of internet dating sites like *Ashley Madison* and *Victoria Milan* that facilitate them (Lamont 2016; see Chapters 6 and 7). Yet, not everyone finds such illicit relationships compelling and some may be put off by the prospect. One blogger who goes on a date via the *Ashley Madison* site reflects that he and his date shared 'similar interests, and to be honest, if she hadn't been you know, married, [he] would have probably have been really excited' *(Jacob, UK, individual blog, 2013)*. Lucy *(UK, individual blog, 2017)* similarly, texts one date about how she thinks 'the Americans have it right when they date a few people at once . . . until you are ready to fully commit to one person'. However, she openly declares that she is 'definitely fucking not converting to polyamory' (2019) and ends a relationship with an openly 'non-monogamous' man.

Nevertheless, despite these uncertainties, we maintain that there is exciting transgression of UK and Australian heteronorms around (serial) monogamy in the practices of daters having multiple profiles and 'dating' more than one person at a time online. Almost all of the accounts indicate that such practices have become routine. For instance, while Jacob was unexcited by the prospect of an affair mentioned above, he later recounts how 'in one busy week [he] dated three lovely Jewish ladies' *(individual internet dating blog, 2013)*. He says that he may have had some moral qualms in doing this, but that dating more than one person at once is 'the norm on online dating'. He continues to do this as is apparent in his later description of having found a more long-term partner, with whom the relationship has become 'proper serious'. He indicates that he intends to tell his other dates about his change of circumstance, but this move does not apparently mean complete exclusivity:

> what about the text message to the Doctor explaining you were seeing someone? Oh, alright, I suppose I should tell you what happened. In between Date 24 and Date 25.
>
> Now, this isn't unusual–I haven't actually closed my account on any of the sites I've joined, including ones I've never got dates through . . . [and a message on one] piqued my interest.
>
> *(Jacob, UK, individual internet dating blog, 2013)*

He claims that having several profiles across different sites is not 'unusual'. It does appear to be a common practice (Couch and Liamputtong 2008: 272; Frohlick and Migliardi 2011: 79). The drift away from previous norms seems incomplete, given the need to explain the now allegedly routine dating of a number of people, perhaps even many, at once. Such shifts in norms become visible, as Elias (2000) notes, because the departure from former practices has to be explained. It is clear from the examples above and the fairly casual mentions that dating more than one person is becoming accepted, as long as only first or possibly second dates are involved. This is the case with Jacob's three dates (with three different women) in a week and also evident when he started 'seeing someone' in between dates 24 and 25. Mundane polyamory is not hidden, indeed he and others proclaim it in their online accounts. Yet, it still gives at least some daters 'moral qualms', even if it seems not to be stigmatised.

This transgression is not typically a radical challenge to monogamy, but more of a temporary departure from it in that the polyamory is most often only acceptable in the early stages of online dating. We have already discussed in our chapter on new norms (Chapter 3), that when internet dating begins to turn into a serious relationship then the key questions become: 'When do we become "exclusive"? When do we have the chat about going offline?' *(Rose, dating site blog, Australia)*. Some (emotional) reflexivity is required to determine the boundaries of this mundane polyamory. Jacob suggests that, for him and his new romantic interest, declaring themselves a couple and her meeting his mum signal the start of a 'serious' relationship. He then texts a doctor he has dated once to let her know that he is no longer available. The return to the normative is completed—although the several profiles seem to still be up. This is a regular feature of mundane polyamory online and suggests that at least some degree of subversion—a more deliberate and conscious challenge to monogamous norms—is encouraged by the apparent availability of potential partners characteristic of online dating.

Possibly more clearly subversive of heteronorms around monogamy is the way that internet dating can provide exciting casual sexual encounters. Subversion involves more considered troubling of the heteronorm than divergence or transgression (Beasley et al. 2012, 2015). The subversion of expectations that hetero-sex occurs within committed relationships is upset by internet sites explicitly dedicated to casual sexual encounters, as well as by instances of such encounters within the wider framework of internet dating (Couch and Liamputtong 2008; Daneback et al. 2007). One male blogger experiences something in between via a site or app called 'Bang with friends', where long-term friendships can be a source for casual hook-ups. This is what happens when he dates a friend who indicates via the app her desire to have sex with him: 'as the evening drew to a close, she asked me to . . . well, like I said at the beginning—dating blog, not a sex blog. Again, use your imagination!' *(Jacob, UK, individual blog, 2013)*.

This coy, Barbara Cartland-style use of the ellipsis in place of sexual description frustrates our wish to know whether the encounter was exciting but is a noteworthy counterpoint to the often pornographic explicitness of many online representations of hetero-sex.[7] We have, for instance, mentioned in Chapter 2 that women and men may use internet dating sites to facilitate casual sexual encounters and some women may blog about their experiences of everyday sex (Muise 2011). However, accounts which document sexual encounters related to internet dating are relatively rare such that the kinds of sex desired or practised by internet daters remain far from clear (DeMasi 2011; although see Fileborn et al. 2015). The very limited descriptions of sex in our sample are generally rather vague and 'vanilla' *(Lucy, UK, individual blog, 2019)*.

That does not mean that the casual sex experienced is not exciting, even if fairly normative. A blogger who slept with one woman he met online on their second date said he 'didn't know sex could be that good. [He had] never experienced such attentiveness or excitement' *(Steven, UK, individual blog, 2015)*. It appears that sometimes women, not only men, may at least try to take up opportunities for sex, even on the first 'date', as one account outlines: '[t]here was the Indonesian girl who clearly had something in mind for me, which didn't include my winning personality. She knew next to no English, leaving us with a failure to communicate' *(Phillip, Australia, online newspaper story, 2011)*.

We are left to imagine that what she had 'in mind' was casual sex, especially given her language difficulties, leaving little chance of communicating verbally. And casual sex can be great for women, as it was for Chiara in her first time meeting with Dante, who she found online and went to meet with the intention of having sex with. After 'four to five hours of talking he finally makes his move'. Chiara explains that,

> Dante puts me at ease though, he says all the right things. Tells me how beautiful and sexy I am, complements me on my red lingerie—yes, I came prepared—and I can't help but enjoy myself. I forget all the bullshit of the last year and have fun like any single 40-year-old should do.
>
> *(Chiara, Australia, guest post on dating blog, 2019)*

This describes, if briefly, a fulfilling casual sexual encounter in a non-pornographic way and from a woman's viewpoint. The account is also subversive of expectations around women's sexuality and heterosexuality more generally.

The subversion of norms of monogamy suggesting that hetero-sex should be reserved for 'serious relationships' is neither new nor inevitably radical, but there is the added element of a seeming subversion of expectations that men initiate sexual encounters, especially casual ones. The men in our sample do not seem to make a habit of using the internet to initiate casual sexual encounters. Indeed, reading further in one account, we see that Steven was hopeful of developing a long-term relationship. He says he 'wanted regular-ish contact' and that they then 'texted every day' *(Steven, UK, individual blog, 2015)*. Nevertheless, the almost 20 years of blog entries by this man indicate that his internet dating relationships seem to be fairly short term, even if they are not more fleeting sexual encounters. Yet, despite the apparent normality of such short-term encounters in online dating, it would seem from the most recent *OkCupid* (2016) report on sex in internet dating that while there has been an increasing tendency for internet daters to expect and have sex soon after meeting, there remain discomforts about heterosexual women discussing sexuality. For example, Lucy's *(UK, individual blog)* candid blog about her dating experiences contains constant references to her discomfort at discussing sex with strangers or on first meeting online or in person. However, via a blog which hides her real-life identity, Lucy does communicate (to strangers) her experiences of casual sex arranged through internet dating sites, despite that discomfort. There are clearly some tensions here.

Some women daters actively engage in casual sex but nevertheless are critical of what they view as concomitant requirements regarding emotional restraint and having to be 'super casual cool girls'. Other UK women note that casual hetero-sex is not normative for women because it is usually deemed to be inevitably oppressive of women, yet insist that it is not necessarily or even revel in the enhanced opportunities for casual sex afforded by internet dating.

> The hardest part is trying to reassure my friends I know what I am doing. When they know it's casual sex they instantly assume I'm being fucked over. When actually I'm aware that whoever it is will not suddenly fall in love with me/want to spend real time with me.
>
> *(Alice, UK*—cited in Gilmore 2019*)*

> Casual sex is just bloody wicked isn't it! I'm very all or nothing, so if I'm not in a relationship I'm having lots of hookups. I'm very proud of having been well 'slutty' in my life because it's great. I cannot stand when people think the only environment in which you can have good sex is in a relationship.
>
> *(Dani, UK*—cited in Gilmore 2019*)*

And, sometimes women even point out that such opportunities for casual sex have been empowering and assisted them in ways that being in a relationship might not.

> You've set the boundaries for why you're there, you're maybe going for a drink first but there's no pretence or confusion. I find myself hooking up with a few people every month, usually a regular casual sex thing . . . It's led to some very fun experiences and has allowed me to explore what I like and don't like, without the pressure of a relationship.
>
> *(Tiffany, UK*—cited in Gilmore 2019*)*

Despite the tensions between risk and excitement in online dating that we have noted in this and previous chapters, it is evident that a focus on the former alone is not the full story. Internet dating can generate innovations in gender relations and one of these arises in the exciting benefits to be gained by women through increased access to casual hetero-sex. As Sally points out, there are drawbacks but the benefits are there nevertheless:

> there were a lot of negatives. It could feel seedy . . . You're trusting people you barely know . . . Most often . . . I didn't have sex at all . . . But Tinder is addictive . . . I sometimes went on three or four dates a week . . . With Tinder, I discovered what it could be to have sex then walk away without a backward glance. That was liberating.
>
> *(Sally, UK*—cited in Moore 2019*)*

The departures suggested by these online accounts are significant in showing how internet dating can involve heterodox forms of gendered heterosexuality.

Emotionally reflexive heterodoxies

Heterodox (non-normative) heterosexualities exist within internet dating and can involve exciting rather than simply fearful forms of risk. In addition, internet dating is not simply about rational, individualised choices and the practices it involves cannot be fully captured by market metaphors. Even where daters might see what they do as akin to shopping, this may be more about shopping as a pleasurable activity than as a form of rational choice. In this and other ways, the experiences recounted in our sample, which are sometimes registered in the associated literature, might prompt novel and less hierarchical forms of gendered relating. This requires further research, but there are indications here of non-normative possibilities at work. Internet dating can provide the impetus for people to navigate competing discourses regarding connection and relationships arising out of their own and other people's emotions. This emotional reflexivity is not guaranteed to undo gender and other forms of social hierarchy, but can, under certain conditions, enable 'lines of flight' from hegemonic norms and allow some movement towards more egalitarian forms of heterosexual relating.

We have traced the non-normative possibilities for heterosexuality, from divergent to transgressive and occasionally subversive and here provide new empirical illustrations of these possibilities taken from our thematic analysis of a sample of UK and Australian accounts of online dating. So far, our analysis of the sample has not highlighted more radical forms of departure from the heternorm, though there are some signs of such radical potential in the literature on the possible links between internet dating and declining racial homogamy. Nevertheless, these accounts suggest that internet dating can excite moves beyond normative discourses amongst the relatively privileged heterosexual groups who undertake internet dating and then reflect upon it in blogs and other sources. Meeting people we desire is exciting and internet dating is exciting because it facilitates that. However, the accounts we analysed also hint at a cycle of excitement and disappointment more specific to internet dating.

The seemingly vast possibilities of meeting attractive others is exciting, but these accounts suggest that this may also involve often boring searching and messaging followed by the disappointing reality of one date after another which fails to live up to the high expectations. The very volume of dates made possible by internet communication may deplete excitement because dates become too ordinary. The excitement of internet dating could promote some divergence, for instance, from homogamy, moving a few of those accounted for here towards finding partners beyond their own class or ethnic background. There are other novelties of interest in these stories such as the (at least) transgressive practice of mundane polyamory. These practices may blur into more deliberate subversion of heteronorms around monogamy because internet dating sites can facilitate casual hetero-sex. However, there is limited discussion of this in the experiences shared in our sample and more research is needed to determine how far casual sex related to internet dating might be subversive in ways that are pleasurable for women and enhance more egalitarian sexual relationships.

Internet dating has the potential to excite shifts beyond the normative. It does not always or inevitably do so and may reproduce gender norms. Fear might be closely linked to a pleasing excitement that moves people beyond what they usually feel is comfortable. Where internet dating offers some transgression, through mundane polyamory, for example, this may start as temporary. The wanderings from the norm of monogamy typically, though not inevitably, appear to stop when internet daters find a person with whom they wish to establish a longer-term relationship. On occasion, departure from the norm may become more subversive, as evident in the ongoing deliberate multiple dating, interspersed with monogamous relationships described in the longstanding blogs of Lucy and Steven. This shows some swivelling from divergence to transgression to subversion and back again (see Thompson 2014: 134; see also Chapter 3, note 1). There are other subversions hinted at around internet dating, such as allowing women as well as men to seek and enjoy casual sex, but we know only a little of users' experiences about this. What is evident is that internet dating is not simply experienced emotionally in terms of fear of risk, it also contains more pleasurably exciting emotional and possibly sexual experiences. The navigation of emotionalised discourses encountered in relational sociality and the actions which ensue, constitutes emotional reflexivity and provides insights into how such reflexivity is key to navigating the complexity of contemporary (hetero)sexual relations. It is also key in understanding how it is possible for such relations to depart from the normative and sometimes be reformed in novel and potentially more mutually respectful and pleasurable ways.

Notes

1 There are ongoing debates concerning the use of terms like 'emotion' and 'affect' with some scholars using these interchangeably and others insisting on a distinction between them (for example, Ahmed 2004; Berlant 2011; Barnwell 2018; Hemmings 2005; Von Scheve 2018). While we see some uses in accounts of a presumed distinction between them, here we have referred to emotions, as alluding to their social/contextual and embodied gestural and visceral character.

2 This form of feminist stance has also been termed sex-positive feminism. The debates between this stance and counterviews (often erroneously described as anti-sex or more usually sex-negative/antipornography feminism) reached their height in the 1980s and are commonly called the feminist sex wars (Bazelon 2015; Comella 2015). Many commentators consider these debates to be based in different socio-political positions associated with Liberal, Foucauldian and Queer theoretical frameworks as against those associated with second-wave radical feminist frameworks respectively.

3 The anthropological term endogamy is also linked to homophily and homogamy but adds a further element of social prescription requiring people to connect with an in-group and sanctions against connections with those outside the group (Hu and Nash 2019).

4 These indications of an increase in heterogamous partnering in Australia do not however signal that racism in online dating is no longer an issue (see Chapter 7; Stokel-Walker 2018).

5 It is worth pointing out that increases in inter-partnering in Australia (and the USA) are modified by gender differences. For example, it is not unusual for women of Thai, Japanese and Philippino ancestry to marry Australian-born men, but not the other way around, while men from Lebanon and Turkey are more likely than their female compatriots to marry the Australian-born. Moreover, while inter-marriage between people of different countries of birth is increasing in Australia, there remains a tendency to partner with those who come from a similar ancestry/region—such as Pakistan and Bangladesh (Khoo et al. 2009; Walker and Heard 2015, 2012). Heterogamous interethnic/interracial partnering is unevenly radical and homogomous elements may remain in what appears heterogamous.

6 According to the Australian site, *RSVP*, a majority of their users of the site (data obtained from an in-house survey of over 2,400 members online in December 2015)

were comfortable with parallel dating and intriguingly more women than men in this report on *RSVP* research were unflurried by the practice:

> 54% of men and 63% of women say it is ok in the early stages of dating to continue dating others. It takes around 3–7 dates for most singles (57%) to think about dating exclusively (RSVP Date of the Nation Report 2016).

Advice provided on dating sites commonly accepts, even embraces, mundane polyamory but only to a point. As this advisor on *RSVP* recommends, parallel dating may be deemed beneficial but not apparently parallel sex. Authoritative commentary is strongly inclined to counsel a return to heteronormative coupledom.

> For some of you, it will be natural to date as many different people as you can while you're in the process of getting to know them . . . you can organise first dates and then meet up . . . At this stage, there's definitely no harm in dating a number of different people. However, where things tend to go wrong, is if you decide to sleep with several of these new love interests that you're dating. This is where I think you need to draw the line. Dating and getting to know numerous singles can be healthy—sleeping with a number of them all at the same time will typically create heartache, confusion and anger (Aiken 2016).

7 This coyness may relate to the blogger being an established journalist writing under his own name and thus not wanting to attract negative attention by publishing overly explicit material. However, other of his posts do describe sexual encounters in more detail. It may therefore be his friendship that restrains him from a 'kiss and tell' story.

Chapter 5

Chemistry

Finkel at al. (2012: 3) ask if the claims of internet dating sites to have altered the dating landscape for the better are justified. The central question which frames this book involves explicating their query concerning evaluation of the benefits of internet dating making it more explicitly socio-political.[1] We interrogate whether heterosexual internet dating as a new technology of intimacy offers opportunities for enhancing social equality, specifically in delimiting the constraints of gendered heteronormativity and advancing gender equality. In that light, we now turn to perhaps the core terrain of the internet dating landscape which we have condensed under the term 'chemistry'. The language of 'chemistry' is employed as a composite vocabulary for examining embodied sensations and emotions as well as practices and events, which draws together both the romantic and the sexual to include everything from initial interest to immersive relationship building and establishment. We make use of this overarching, and potentially flabby and inexplicit term, because this is how internet daters express their understanding of what is crucially at stake when they date. All our investigations in previous chapters of norms, narratives and emotions/excitement as well as the myriad other aspects of online dating, are almost window dressing compared with what most daters take to be *the* moment of dating *per se*.

We employ 'chemistry', as an inclusive portmanteau terminology which for us enables exploration of a spectrum of sensory, emotional and practical experiences, particularly involving the romantic and/or the sexual (though not entirely exclusively). In this we attempt to mark out a space in internet dating which is broadly inclusive. Here, we differ from many writers who are inclined to focus on specific aspects of what we have included under this eclectic rubric, such as attractiveness, spark (which tends to be used by those in our sample as referring to initial sexual attraction to an online profile or arising in early online meetings), a sense of affinity, flirtation, desire, fantasy, (ongoing) sexual attraction, romance, seduction, compatibility, early relationship building—the honeymoon or courtship period—and intimations of romantic love.

However, before considering these aspects as connected, it is worth noting that some of them require a little more discussion, as individually they are not entirely straightforward entities. For example, 'attractiveness' may be associated with what Hakim controversially describes as 'erotic capital' and Bruch and Newman conceive as competitive 'desirability' (Bruch and Newman 2018; Hakim 2012; see Green 2013 for an alternative approach). By comparison, as noted in earlier chapters, we are wary of employing terms or characterising terms in ways which foreground individualised, competitive, market-based metaphors for dating experiences and practices.[2] Hence, attractiveness is one of a number of labels concerning erotic life that we include under 'chemistry' which we think require caution. Its use can lead down analytical paths with which we have some quarrels.

Whether flirtation should be included in the list of aspects of 'chemistry' is also capable of being disputed since some commentators attest that flirtation is, unlike seduction, not a means to an end in that it does not have the goal of sex with the other person but rather is an end in itself. In similar fashion, flirtation is differentiated from seduction's entanglements with power and is seen as not requiring anyone to 'surrender'. We, however, are not entirely convinced by these approaches. If flirtation were indeed without any erotic/romantic goal, it would not be relevant to our discussion of 'chemistry' in online dating which has at least some goal. By contrast, we are inclined to suggest that flirtation entails a spectrum rather than a singularity of expression or perhaps occupies what Mølbak describes as a liminal space. In these ways it cannot be entirely separated from goal-orientated practices like seduction in our view and is by no means entirely free of investments, including investments in power. For example, the #MeToo movement, among many other instances, reveals varied understandings of flirtation associated with gendered power relations. Hence, we share a sense of the equivocal character of flirtation: 'flirtation would not be itself if it did not hold out the possibility, however distantly, of going beyond mere flirtation' (Hoffman-Schwaartz, 2015: 15). This account of flirtation, in our view, enables it to be definitively included in analysis of 'chemistry' in internet dating (Bartlett et al. 2019; Fleming 2015; Hoffman-Schwartz 2015; Hoffman Schwartz et al. 2015; Mølbak 2010). We begin that analysis by looking at the central importance of chemistry and exploring its features.

Beginnings, value, technology and timing

It can be said that for the vast majority the whole of internet dating is directed towards and about the moment of experiencing 'chemistry', which typically in the first instance takes the form of a 'spark' of attraction. As one woman writing on the blog for a large commercial UK dating site explains, this experience is for her the aim of the process of internet dating:

> I choose to contact the ones who I feel could be compatible to me and then when we meet up, it's a case of seeing whether that elusive 'spark'

> happens. I've not yet met someone who is similar in wavelength to me that I also 'spark' over, but I know it will happen.
>
> *(Sula, UK, dating site blog, no date)*

The chemistry may not be mutual and may not be simply about sexual attraction.

> She kissed me quite intensely for a while, which was more than I was expecting and it was as if to say 'maybe there is something there' . . . I kissed her again and said goodbye for now. Whilst I didn't really see a future for Natasha and I, I did enjoy my trip to Moscow, I probably would have never gone otherwise.
>
> *(Peter, UK, individual blog, 2005)*

Nevertheless, whether chemistry is shared or strongly sexual, or even produces a lasting result, it is the 'deal-breaker' for many an internet dater, which serves to reinforce its vital significance. Steven shares two separate experiences of chemistry.

> Sadly, the fourth date never happened. I texted her a couple of days later and heard nothing then received a call asking me to ring her back for a chat. I did so, fearing the worst, and she said she just wanted to be friends as she didn't feel that there was a spark between us.
>
> *(Steven, UK, individual blog, 2010)*

On this occasion, his feelings of attraction seem not to be shared. However, he returns to the dating fray shortly after.

> [On my] first date of 2009 [sic, 2010?] . . . things couldn't have got off to a better start . . . it turned out to be the most perfect first date I'd ever had. I knew as soon as I saw her that I fancied her and to say that there was a spark between us was an understatement. It was a purely magical evening . . . The next morning . . . we both went our separate ways.
>
> *(Steven, UK, individual blog, 2010)*

Chemistry, frequently initially discerned in the form of a mutually experienced 'spark', is *the* means of discerning the value of the date, but does not always mean living happily ever after. In the above case they had two or three more wonderful dates and then never saw each other again. More typically, stories tell of what happens when there is a lack of spark, signaling that chemistry is unlikely: 'I imagine a lot of online dates end like this—the person is nice enough, but there's no real spark. At least it was a nice evening, and hey, no wounds or infections. So, a win' *(Jacob, UK, individual blog, 2013)*.

Hence, a commenter on an Australian online newspaper article about internet dating responds to the article with advice about how to proceed and what to look for when seeking chemistry: 'My advice to those considering [internet dating] is if you think you may like someone and there's common ground, don't spend hours talking. A meeting will cut to the chase and determine if there's spark and attraction and compatibility' *(Commenter on Phillip's story, Australia, online newspaper story, 2011)*.

Similarly, Andy, a middle-aged Scottish Londoner who has been on *Tinder* for a number of years, offers advice which reiterates that the quintessence of internet dating is chemistry:

> [y]ou create this perfect profile, using your best photographs and most creative lines of text, to create this persona, and you start to believe this persona that you've created. Everyone else on the app is doing exactly the same thing.
>
> You spend a few weeks sending lines of text to each other, and you eventually arrange to meet. At the meeting, it's instantly recognisable that neither of you can live up to the expectations and you part company.
>
> If there's no chemistry, there's no chemistry.
>
> *(Andy, UK, online news story, 2016)*

Ironically, the core to this online technology of intimacy is almost always put to the test, resolved, offline. If our discussion of narratives, norms and emotions/excitement in Chapters 2 to 4 relate more to the organisational preparation for offline practice, chemistry is certainly fuelled online but for most daters is crucially about the offline moment. Indeed, it could be said that offline chemistry is pivotal to the entire paraphernalia of online dating. This makes online dating somewhat different from many, if not most, other forms of online social activity and aligns it more with other forms of dating. Internet dating appears to be rather a species of dating with a particular front-end mechanism, rather than an activity that is sufficient *within* its online location. That suggests that internet dating may well not offer anything especially new in the way of social innovation around heteronormative gender relations and gender justice, its specificity being more of a technical issue of form than a matter of changed content.

Chemistry is not necessarily revolutionised by internet dating. Indeed, the tendency towards replication of existing social norms regarding gender may be furthered by the front-end technological aspect of internet dating. In this context, Tannenbaum et al. (2019) have noted, for example, algorithmic bias which gathers up and perpetuates existing norms. Regressive reiteration may even occur when the technology appears at a surface level to be responsive to innovations in gendering. For example, Facebook's upgrading of gender identification options from 2 to 58 in 2014 suggests considerable room for such innovations. However, Bivens asserts that the database at its deeper level actually reconfigures non-binary users into a binary system (2017). In short, there is relationship between chemistry and technology in online dating that deserves some clarification. We suggest in this context that, while the crucial 'chemistry' driver of internet dating almost invariably arises offline, the online technical element of internet dating is by no means neutral and contributes to the shaping of heteronormative gender relations throughout the dating process. Although this front-end technical element can certainly be seen to reproduce and even enhance gendered power differentials, it is also possible that it can contribute to innovative developments. For instance, for all the criticisms that might be made of the structure of *Tinder* as an appearance-based and male-dominated hook-up site, the *Tinder* update of gender identity terms in 2016 to include 37 additional options appears to offer some opportunities for new directions in gender relations (Mallenbaum 2016).[3]

In sum, the key significance of chemistry in internet dating, while usually triggered online, nevertheless gives weight to the offline moment. This would appear to signal that gendered social norms regarding heterosexuality are, despite their comparatively newfangled technological beginnings, likely to be maintained in internet dating. Yet, the technological beginnings of internet dating do also have some contributory effect upon the development of offline chemistry. The effect can be shown to maintain heterosexual gendered social norms, but this is not the whole story. For this reason, we wish to look at the framing and accounts of chemistry for evidence of non-normative possibilities.

Connection: Chemistry's core

What bundles the several aspects mentioned earlier together under the rubric of chemistry for our purposes is that it refers to a sense of magnetic connection. For the great majority, this connection is concerned in some way with sexual interest, with erotic life, even if that sexual interest is conceived in romantic rather than strictly erotic terms. Yet, as we note late in this chapter, there are also a few outlier characterisations of chemistry in terms of emotional fondness which minimise or eschew the sexual. Whether the term is employed in its predominant form to refer to sexual interest or its relatively much less common form to refer to non-sexual emotional bonding, chemistry importantly in some sense calls attention to *beginnings and immediacy*. It is not usually a language of long-term, ongoing and established intimate partnership.[4] This is why it is such an important word for internet dating. Users are necessarily constantly engaged in attempting to assess possibilities from the earliest point of contact with potential dates. Once they establish partnerships offline, online dating (almost always) ends and the language of chemistry typically becomes retrospective—that is, about happy beginnings. It is in the anticipatory interval that chemistry comes into view. While chemistry is not always mutual and can arise in the online preparatory stage, it is usually strongly associated with initial meeting offline. As Allysa notes in relation to video chats that became the necessary replacement for such meetings during the Covid-19 virus pandemic, chemistry is not particularly cognitive but rather linked to visceral embodiment. Video chatting, she says, '[is] not the same. It's the chemistry, you can bounce off each other when you're face to face. Even a touch. A touch is a huge thing in dating (*Allysa, Australia*—cited in Leabeater 2020).

It is necessary to point out in this setting that we reject notions of a clear division between 'the virtual and the real' and of any distinct separation between online and offline. Our approach acknowledges that online interactions can be viewed as another aspect of everyday offline relationships. However, we also note that users of internet technologies, including younger users, clearly distinguish between on and offline (Miguel 2018: 6; Miller et al. 2016: x, xiii, 7, 112). Moreover, we do not accept that these modes are interchangeable or simply equivalent. Nor does the language of 'modality switching' (Ramirez et al. 2015) seem to capture this non-equivalence. It is not simply a question of technical modality. In relation to dating, we therefore also wish to highlight the specificity of online and offline to bring attention to the specific emotionally charged and often erotic, but importantly temporal location of chemistry in the arc of developing intimate relationships.

While chemistry may not be mutual nor necessarily about sexual attraction, in the daters' narratives the supreme form of chemistry entails mutuality—typically a dyadic connection—even in the face of significant barriers:

> when I met him, it was just so easy and great. We kissed at the end of the night. When I got home, he texted me that he was deleting his account, and I was like, OK.
>
> This is actually my first serious relationship, and we're at almost a year now. Things are great; I'm so happy. The one thing that is a bit of a problem is, funnily enough, although we're both Indian, he's Muslim, and my background is Sikh. Our families don't know about us, because it's kind of a taboo. We're thinking about how and when we'll tell our families.
>
> (*Rani, UK*—cited in Oswaks 2015)

In the narratives of our sample of internet daters, chemistry most powerfully figures not only as a shared recognition of connection but furthermore as a mutual sense of being apparently unusually, even uniquely, drawn to the other. Chemistry may prove in time not to generate an ongoing and significant intimate relationship, but it most often asserts that possibility. This kind of narrative is a staple feature of the advertising offered by internet dating websites to showcase 'success stories'. For example, on a large Australian site, a 'single mum' describes her experience of a virtually instant sense of connection.

> I joined [dating site] in September 2012. As a single mum working 2 jobs and no desire to go out, friends convinced me to try it and see what happens. I joined. Within 5 minuets I received a message (sic) . . . It was Jackson. We started messaging back and fourth, than he asked for my number

(sic) . . . that Friday morning, I picked him up and we have been inseparable since. I met my soulmate! March 17th, 2018 I married my soulmate!!
(Connie, Australia, dating website commentary, 2018)

Sex, intimacy and 'social flesh'

With this characterisation of chemistry in mind as primarily a language of dawning connection, of erotic beginnings—usually mutual—offering the promise of consequential intimacy, it becomes necessary to explore the role of sex and sexuality. What do sex and sexuality mean in this context? While chemistry is not simply a question of sexual attraction and desire, it does appear most often to require some level of sexual interest and typically sexual connection. On this ground, we note the significance of sexuality in internet dating, intimacy and social life more generally. As we will discuss in more detail in the next chapter (Chapter 6), the link between sexual health and well-being is now well established. *The Global Study of Sexual Attitudes and Behaviours*, which involved a large sample of men and women aged 40–80 from 29 countries,[5] found that over three quarters of participants agreed that 'satisfactory sex' was essential to maintenance of relationships and was associated with an overall sense of health and well-being (see for example, Laumann et al. 2005). The psycho-social and physical benefits of sexual health and well-being include among many others, improving stress reduction and sustainability of relationships, along with reduced risk of coronary events such as stroke (Planned Parenthood Federation of America 2007; World Health Organisation 2006). As Debrot et al. (2017) note, sex is strongly associated with overall well-being and has relational benefits in that it is linked with embodied affectionate connections between partners (see also Whitbourne 2017). Conversely, loneliness in the face of lack of social/sexual connection with others is an emotional, psychological and social problem, as well as a health risk (Cacioppo et al. 2002; Holt-Lunstad et al. 2017; Owens and Sirois 2019; Swinburne University 2016; Victorian Department of Health 2017). Apart from lack of social/sexual connections being linked to sexual risk-taking behaviour, loneliness is a risk factor for heart disease and depression. As Andy, a seasoned internet dater, points out, being without a partner can be a driving imperative to continue with internet dating, despite many disappointments: '[t]he whole process is very addictive . . . The addictive part tends to come mainly from the loneliness, which is the whole reason that we're on this' *(Andy, UK, online news story, 2016)*.

Given the advantages attached to intimate connections with others and the obverse disadvantages associated with their lack, perhaps it is no wonder, that daters place a high value on finding the ineffable, the 'chemistry' that might enable the beginning of connection and even an intimate sexual relationship that might last.

The evident importance of experiencing chemistry and its possibilities indicates that analysis of internet dating—as a growing, indeed unprecedented arena worldwide for seeking intimate relationships—is a topic of profound public consequence. The chemistry sought by internet daters is linked, as Finkel et al. (2012) point out, to the central role intimate relationships play in our physical and emotional well-being such that connection with others has been depicted as a 'fundamental human motivation' (Baumeister and Leary 1995). Beasley and Bacchi underscore the visceral interdependent character of human sociality by referring to 'social flesh' to evoke its power in human lives (2012, 2007, 2000; see also Chapter 4). This approach is highly relevant to considerations of chemistry in internet dating because, although scholarly literature on personal relationships is comparatively new and undeveloped (Finkel et al. 2012: 3; see also Reis 2012), the term social flesh offers a critical perspective on the market-oriented emphasis on rational strategic individuals that we have already noted in much of the existing literature on partnering online. In this context, Finkel et al. (2012) remind us of the concrete implications of the power of social flesh in that the failure to find intimate connections and associated loneliness is a significant social problem linked to mental and physical ill health and economic costs (see further discussion in Chapter 6; Fakoya et al. 2020; Ferguson 2011), whereas intimate relationships produce better health and quicker recovery from illness, as well as extending longevity (Holt-Lunstad et al. 2010). In a somewhat less instrumental vein, Finkel at al. draw attention to wide-ranging research indicating that the presence of such a relationship provides 'one of the strongest predictors of happiness and emotional well-being that has been measured' (Finkel et al. 2012: 3).

The rise of internet dating, and the lessening of social stigma attached to it, would seem to be a highly promising development if it facilitates satisfying relationships and makes it easier for ever larger numbers of people around the world to have the opportunity to date (YouGov 2017b). The emphasis upon benefits in this approach enables a more positive engagement with online dating, chemistry and sex than those perspectives which, as we have noted in previous chapters, stress fear, risk and danger. The scholarly literature dealing with online dating is frequently inclined to concentrate on issues like dishonesty/fraud, sexually transmitted infections, harassment and violence, whether the research focus is upon youth or older people (Beasley and Holmes 2016; 207; Couch and Liamputtong 2008; Lykens et al. 2019). However, even the apparently more 'pro-sex' writings on internet dating which embrace opportunities for sex and romance,[6] still retain a strong thread of risk management and health warnings while outlining its advantages for seeking intimate partners. As Lykens (2019) tellingly notes, despite a range of studies outlining young peoples' use of online spaces, little is known about their use for sexual and romantic exploration. Even the literature which describes internet dating in terms of potential connection—rather than zeroing in on social and health risks and/or other dangers—usually fails to consider it as an exciting and enjoyable activity (see also Chapter 4). The involvement of friends, family and community and experiences of curiosity, pleasure, playful flirtation and intimacy are ignored. The striking emphasis on utilitarian health aspects when it comes to research on older people amounts to neglect of considering dating in terms of sociality, of social flesh. Important dimensions of older daters' experience thereby become invisible. Although studies on young people and internet dating do not exactly provide much in the way of understanding it as a meaningful social activity, this is even less evident in work dealing with older daters, such that critical aspects of older people's everyday lives are discounted and ageist conceptions of older people as desexualised and passionless are reinforced (see Chapter 6).[7] In short, there is a decided shortage of studies dealing with chemistry in internet dating and what limited attention is given to its embodied emotional and social elements is largely based on rather meagre accounts of the experiences of young people—and indeed very frequently of young Americans.

By contrast, we want to foreground chemistry in order to evoke more expansive and nuanced understandings of dating experiences, including registering its strong connotations of hope, imagination and fantasy. Chemistry is a terminology which captures a very human embodied emotionality concerning intimacy and connection that is not reducible to investment or market-based strategies to arrive at a 'good deal', or supposed elemental reproductive/mating drives of the species arising from biological evolution or utilitarian concerns with avoidance of risk. The term and its crucial significance in the voices of daters within our sample and in other sources, highlights our sense of what is missing in much of the existing literature on internet dating—a gap which hints at the trivialisation of social flesh, emotionality, intimations of desire and the ineffability of human intimate relationships. There is indeed often a strange avoidance in the literature of a sympathetic response to daters and their concerns, hidden in an auditing orientation and in the common resort to objectivism and technicism. Our aim is instead to bring the voices of daters and their longings and dreams to the centre stage. Chemistry speaks of potential promise. It entails a moment, not only of calculation and caution, but also one of possible poetry.

The poetics of hitting it off and getting on

It may not always be expressed in conventionally poetic ways, but many daters provide descriptions of mutual chemistry leading to an ongoing relationship. In consequence, when internet dating is subjected to criticisms, it is common for users to rise to its defence and in so doing point out their experiences of chemistry. For example, when an Australian blogger asserts that dating sites are 'bad news' because they are full of scammers and 'fruitcakes', and even speculates that most of the profiles of women are 'made-up' to trap men into signing up *(Eric, Australia, discussion forum post, 2007)*, his cynicism immediately leads to many responses which counter his perspective. The responses from male daters dispute this negative account. Here are a few exemplary instances.

> I met my partner on RSVP almost 2 and a half years ago, we were both trying it for the first time, just for a laugh mainly and we just hit it off straight away and once we met we had an amazing first date and got on like a house on fire.
>
> *(Michael, Australia, discussion forum post, 2007)*

> [D]ating sites have their place in the world . . . in saying that (sic). i met my wife 4 years ago. One rather interesting thing was is (sic) that we lived only 4 streets away from each other in the same suburb . . . We have been married for 2 years and have a healthy and happy 9 month old boy!
>
> *(Sam, Australia, discussion forum post, 2007)*

> Met my missus on [dating service] site. Met for drinks. Totally clicked. Havn't left each others side since (sic).
>
> *(Flin, Australia, discussion forum post, 2007)*

Women's accounts of mutual forms of chemistry in our sample appear to be just as concise and mysterious about why they 'hit it off' or 'got on' or 'clicked'. Belinda met David on the large Australian site, *RSVP*, which has a generally older cohort of users than many dating websites and apps (most members are 35+—Sheftalovich and Smith 2020a). In 2014 they were married and expecting their first child. Belinda had made use of two dating sites over a period of 18 months prior to meeting David and reports that she went on hundreds of dates: 'I had dates with doctors, lawyers, a millionaire, but I had no feelings for them apart from friendship. I knew the moment I saw David: "This is the one." I'd never had that before' *(Belinda, Australia*—cited in Sheftalovich 2014*)*.

Belinda just knew, or felt, that David was 'the one'. As these descriptions indicate, chemistry itself is very often not delineated further than simply using the word itself or its analogues, typically swiftly followed by 'proving' its existence via noting 'success' in forming a connection/relationship. Chemistry is critical to online dating but, like sex, accounts of it are usually short on detail. Its parameters, its explanation, appear imprecise and remarkably inarticulate, perhaps because it is felt, is seemingly primarily subconscious and embodied, and hence largely wordless. Chemistry may produce poetry, but it is more in the form of haiku than sonnet. The seeming difficulty of finding more than a few words to put flesh on the bones of chemistry is in inverse proportion to its importance.

The dark side of chemistry

However, as we have suggested already, recognition of chemistry's profound significance and foregrounding the pleasures and 'successes' of internet dating does not preclude caution about singularly positive 'pro-sex' agendas in relation to internet dating. It is vital to remember that while loneliness is a serious individual and social issue, being alone (solitude) should not be equated with it (Jamieson and Simpson 2013). Relatedly, although sex is linked with many benefits, it cannot be assumed that sexual partnerships, sexual activity or sexual interest are essential to all. There are significant problems associated with the obligation to be sexual and indeed to be (hetero)sexual in gender normative ways. Chemistry may be capable of inducing poetic rhapsody but it can also have a distinct sting in the tail.[8] The endless reiteration of chemistry as central to internet dating—as the crucial key to fulfilling the promise of 'heterohappiness' (Marshall 2018) and its corollary, the supposedly necessary 'natural' unit of hetero-coupledom—is suffused with heteronormative underpinnings. Adrienne Rich's 1980 account of 'compulsory heterosexuality' remains relevant here. Rich argues that if heterosexuality were truly as naturally and normally widespread as is usually claimed, there would be no need for the material and symbolic punishments directed at those who fail or refuse to comply. We note here that the way in which chemistry is centralised in internet dating may potentially constitute singles as pitiable and unhappy or immature and exclude other possibilities. The Janus face of the heterosexual imaginary is manifest in chemistry's associations with playfulness, joy and pleasure *and* the fear and anxiety attached to loneliness, abandonment, or being unloveable (Beasley and Brook 2019: 150–151, 155). And, unfortunately, while this 'dark side' of looking for chemistry online is ever present, it is also unmentionable, as one inveterate dater acknowledges: 'Online dating may have (sort of) solved the supply challenges of romance, but . . . [i]t's still the case that nothing is less socially acceptable than admitting you're lonely and longing to be loved *(Sarah, UK, online newspaper story, 2015).*

The inclination to respond to the siren call of chemistry as a kind of holy grail that all internet daters should seek is intimately bound up with a demand to be normatively gendered and (hetero)sexual. Thus, while we have outlined the problems of an overly risk-oriented approach to online dating, nevertheless an undiluted 'pro-sex' orientation ignores ongoing systemic inequalities. While we object to a doom and gloom focus on risk, we do not lose sight of the dangers and difficulties of internet dating, particularly for heterosexual women and with regard to gendered power relations. On the latter score there is much to support a case for internet dating upholding normative heterosexual gender relations and not advancing opportunities for gender equality. In this context, we have noted so far that assertions of gendered sexual dominance and control by heterosexual men continue to plague internet dating. We see also the reiteration of a variety of other normative practices, such as heterosexual men looking for casual sex while not disclosing this and viewing dating as a game to achieve this outcome, placing considerable emphasis upon women's physical appearance, and stating preferences for women markedly younger than themselves. In addition, we have pointed out that heterosexual men, and more rarely heterosexual women, make use of internet dating to cheat on and hence deceive their existing partners. Furthermore, our sample and associated literature shows heterosexual women's inclination to focus upon men's financial and other status measures and their expectation that men should undertake most of the work required to establish intimacy, as well as usually bearing the financial cost of this. And finally, the force of hetero coupledom looms large for both heterosexual women and men. This all sounds rather dismal and scarcely suggests it's time to pop the cork on the celebratory champagne. Yet so far, we have asserted that there are signs of heterodoxical innovations, ranging from minor departures associated with divergence to temporary transgressions and occasionally deliberate—though perhaps unassuming rather than sensational—subversive practices.

Heterodox chemistries

We now look more closely at heterodox possibilities specifically in relation to chemistry. We consider here what might be constituted as rather more radical forms of novel practices which include more challenging varieties of subversion, even blurring into intentional dissidence. Dissidence in this case may not problematise the boundaries of heterosexuality with regard to the homosexual/heterosexual binary but may nevertheless gently disturb other normative boundaries such as gender subjectivity around hetero-masculinity/hetero-femininity and, significantly, may trouble the heteronormative authority of heterosexual sex and heterosexual coupledom.

Some shifts from gender normativity are evident in relation to women's search for chemistry, as we have suggested in previous chapters and earlier in this one; for example, increasingly instigating dating contacts, taking charge of aspects of the dating process, taking up casual sex and becoming somewhat less attentive to men's income.[9] McWilliams and Barrett also point out signs of older women becoming more proactive in terms of seeking their own interests and maintaining their autonomy, as well as shifting from the 'scripts' of past relationship experiences in favour of more egalitarian and sometimes less involved partnerships (2014: 413–415, 423).

When we turn to men there are also some signs of departures from gender normativity which have implications for the development of heterosexual chemistry. Many studies of internet dating reiterate gender stereotypical perspectives on sexual attraction/connection promulgated by both men and women daters in terms of men preferring younger women and women indicating a partiality for older men. However, there are some notable departures in our research and some within the scholarly literature which indicate that such gender normativity regarding age disparity as necessary for chemistry cannot be viewed as entirely given. Apart from our sample of sources which provide opportunities for us to access daters' experiences, we have also undertaken—with two other researchers—a cross-national qualitative content analysis of internet dating sites and dating profiles (see Chapter 7). Heterosexual men in our sample of UK and Australian dating profiles consistently put forward gender normative idealisations regarding meeting attractive women younger than themselves, in that they express in their profiles a preferred age range for potential partners which includes women younger, often significantly younger, than themselves. A 20-year age gap is not unusual. By contrast, women in our profile sample, tend to nominate age ranges for potential partners around their own age.[10] Yet, before jumping to any straightforward assumptions about the ongoing force of gender normativity and inequality, it is important to note that in practice these heterosexual men apparently found women of around their own age appealing enough to go on dates with them. Indeed, it would seem in the wider literature that heterosexual male daters at all ages do not restrict themselves to significantly younger women. Kelley and Malouf's study of hypotheses concerning gender and 'mating' preferences is telling in this context. They point out that most studies of internet dating reiterate the view that heterosexual men prefer younger women and that indeed in their investigation of newspaper matchmaking the matchmakers assumed this and thus actually lined up men with younger women. Yet, the participants' ratings of their arranged blind dates did not conform to this assumption of a predilection for gendered age disparity. In short, participants showed no indication that age had any effect on their assessment of the 'success' of their date (Kelley and Malouf 2013: 5, 7). The daters did not apparently find that age was especially important in finding connection.

In short, normative assumptions and even declared preferences regarding intimate relationships are not exactly the same as chemistry in practice, let alone necessarily productive of a 'successful' match. While men do mostly marry women younger than themselves the world over, nevertheless the vast majority of men in Western countries are likely to partner with a woman of a similar age. In these countries only 'about 8% of all married heterosexual couples can be classified as having what might be designated a large age gap (ten years or more)' (Karantzas 2018).[11] For instance, in 2006 around three quarters of couples in Australia were aged within five years of each other (ABS 2009). In 2011, the average age gap between Australian men and women in heterosexual couples was 3.7 years (ABS 2013).[12] In the USA, UK and Australia marriages with a significant age disparity between partners are not only outliers but often attract social disapproval (Sela et al. 2018), expressed in the use of a variety of derogatory terms even when the older partner is male and hence the partnership is stereotypically gender normative. To put this perhaps more bluntly, heteronormative 'ideology' or prescription does not equal living practice (Holmes, 2011; Seidman 1991: 6), in gender or sexuality any more than in anything else (see also Chapter 4). We do not wish to ignore here the significance of social assumptions and declared preferences for conceptions of heterosexual partnership between older men and younger women and the implications of these assumptions and preferences for gendered power relations. It is worth noting in this context that countries which have the biggest age discrepancy between heterosexual couples also have significant gender inequality (Ford 2019).[13] Yet, it is also important not to presume that social norms and stated beliefs are a sufficient guide to social change over time. Dinh et al. have pointed out in this context that, 'despite the wealth of insight user-generated data online dating has revealed about latent and stated mate preferences, there remains significant uncertainty regarding the ways these preferences have evolved over time' (2018: 3).

We would add that there is little in the way of analysis comparing latent, stated and in-practice preferences and almost no analysis of the ways in which in-practice predilections might have changed over time. Relatedly, we note a gap in the literature regarding consideration of whether these different levels of preference (that is, what people might appear to want, what they say they want and their actual practices) are inevitably in synch, particularly over time. It is possible that they might not always be in synch, and perhaps increasingly might not be. Chemistry in practice may confound gender normativity.

Heterodoxy and troubling chemistry?

And now we turn to what might be considered more radical innovations with regard to heterodoxical heterosexualities that challenge the importance and/or understandings of chemistry. These innovations, which suggest somewhat more confronting forms of subversion and even gesture towards intentional dissidence, disturb the heteronormative centrality of heterosexual sex and heterosexual coupledom. The commanding even supreme place in internet dating of chemistry—understood as magnetic connection founded in sexual interest—in such innovative moments may be *undone, deposed and/or completely redefined.*

We wish here to draw attention to several features of heterosexual internet dating practices which might indicate innovations that can upset the reiterations of gendered subject norms often associated with the supposed centrality of chemistry in begetting heterosexuality and hetero-coupledom. In particular we refer to two main practices: *evading or delaying contact* and having *other purposes/aims* than developing heterosexual relationships. This involves a re-thinking of certain widely recognised interactions between women and men in online dating. We begin this re-consideration by looking at women daters' practices in light of several features of heterosexual men's dating. We note here the marked inclination of heterosexual men to initiate first contact with women and to do so often in an unselective way with limited regard to compatibility (Dinh et al. 2018:5; Tyson et al. 2016), the tendency of these men to re-partner more quickly than women after a relationship ends or a partner dies (McWilliams and Barrett 2014: 415), their seemingly widespread dissatisfaction with the number of responses they receive from women, at least in the

USA (Pew Research Centre 2020),[14] and finally their common complaint that women's failure to respond or delayed response to their initiation of contact shows that these women are not seeking chemistry and are not open to relationships. With regard to the latter point, as a middle-aged divorced woman who provides a regular online newspaper column on internet dating points out, men often express frustration concerning women's comparative unresponsiveness and some explain this in terms of women seeking their own personal affirmation and avoiding sex.

> Alex is one of the few men I've come across who wants to spend time establishing a friendship via email before meeting. Generally it's a thing that men complain women want. I've seen some man-chat on the internet about it. Some commenters diagnose a need for a lot of attention. For many women, they say, online dating is a vanity project, in which we're trying to get six candidates to email us simultaneously, thereby revitalising our self esteem, without needing to go to the bother of going to bed with any of them.
>
> *(Stephanie, UK, online newspaper column on internet dating, 2016)*

We challenge this linking of women's inclination to not respond or delay responding to men online to stereotypic feminine vanity and sexual game-playing. Many researchers have pointed out other factors to take into account, such as the discrepancy between the numbers of men and women online perhaps encouraging men to be more assertive to increase their chances and women to opt for a more passive role since there is less reason to reply when they are faced with an over-supply of contacts. However, we suggest there may well also be less gender normative or demographic explanations. It is possible that, through evasion or delaying of contact, women meet other purposes/aims that offer some challenge to heteronormativity. In evasion or delay women may not just avoid risk but gain opportunities for sexual and other pleasures that can be understood more broadly than sexual acts, without the costs of emotional/romantic/sexual connection or involvement (see also chapter 4). There may be opportunities in internet dating, during the online stage particularly, for fantasy, imagination, chatty banter, flirting, seduction, and for embodied pleasures in uncertainty. Pleasures may arise with potential partners who are perhaps not only strangers but unlike oneself, and even 'nutters' may have their charms (see chapter 4 on heterogamy and chapter 2). Prolonging the online stage may for women produce a thrill without disappointment. This is not to be reduced to feminine normativity or to women manipulating or 'toying' with men. It may be a means to evade or delay men's inclination to pressure women to have sex and/or partner up, to evade or delay the rather less exciting scenarios of men's off-line power to control face-to-face interactions with women and to evade or delay 'real life' gender inequality. Yet, importantly, the seemingly pleasurable, even addictive practices of evasion and delay are not only enjoyed by women,[15] nor are they only limited to prolonging of online interaction.

In the scholarly literature on internet dating there are indications that many users, both women and men, are not employing dating sites and apps solely for the purpose of seeking sex or a relationship based on the chemistry of sexual connection. In other words, many users are not conforming to heteronormative assumptions regarding the usage and purposes/aims of online dating. For instance, *Tinder*—usually conceived as primarily about sexual chemistry and sexual connection, as well as less notably about the development of intimate relationships—turns out to not be all that much about hetero-sex or hetero-coupledom. A study in 2015 of *Tinder* users in the United States discovered that just over 70 per cent had never met up with anyone through the app. And, in case this might simply be attributed to the failings of the app, the study pointed out that only about 26 per cent declared they were actually looking for sex or a relationship, while a hefty 73+ per cent ticked the box stating that they were 'procrastinating' and using matches to boost their personal confidence with no intention of meeting anyone, or had other unspecified reasons (probably friendship and curiosity, or playing it like a game) for being on the app (Brown 2017, Fottrell 2017; see also Lutz and Ranzini 2017).

The researchers in this 2015 study of American *Tinder* users seem to imply that the failure to set up offline dates means these users are somehow indulgent, indolent, or under-confident in terms of the presumed objective of sexual/romantic chemistry. Indeed, the behaviour of the mostly young college students in the study is intriguingly labelled in rather derogatory ways by researchers and other commentators. When the participants were asked why they were on the *Tinder* app, the choice associated with not looking for sex or a relationship was named in the study itself as 'procrastination'. Other commentators further describe their avoidance of dating as voyeurism, defensiveness, being picky and judgmental, perfectionism and wasteful daydreaming which might impact upon their mental health, result in social isolation and actually prevent them from developing intimate relationships. A Communications academic, for example, confidently asserts that 'swiping endless photos and fantasizing about each one' is 'not conducive to forming a good match, and it's not exactly a productive use of your time'. As for meeting a potential partner, it 'decreases your chances of doing so' (Fottrell 2017; see also Vine 2018). Once again, there is a distinct reminder of Adrienne Rich's account of 'compulsory heterosexuality'.

Instead, we propose that, whatever the risks of dating apps like *Tinder*, it is also possible that young women and men on such apps might be using the possibilities of this dating technology to do what we have described in relation to women daters on dating apps and websites *per se*. The pleasures associated with *evasion and delaying of contact* bring into view *purposes/aims that are not strictly to do with hetero-sex or hetero-coupledom*—even though they arise in a setting that precisely offers the promise of sex and coupling. Indeed, it would seem that there is some evidence here of deferrals, postponements or refusals of heteronormativity which might be termed heterodoxical subversion in that these practices are intentional and ongoing, though probably not permanent nor very deliberately aimed at upsetting the heteronorm. Yet, despite this apparent departure from the requirements of the heteronorm, there may still be some chemistry here regarding potential partners, even if not chemistry in the mutual sense. In this setting chemistry can be reconfigured as more concerned with elements like titillation and fantasy flirtation, retaining perhaps immediacy of connection for the dater but not necessarily about the beginnings of a relationship with a potential partner. Here, chemistry's common association with the mutual inception of intimacy with another person, is expanded beyond the romantic/sexual agenda typically expressed in earlier narratives outlined in this chapter. Rather than chemistry referring to mutuality or even to a dater's sense of immediate interest in a possible partner, chemistry in this case calls attention to a dater's experience of something of the frisson of dating but with a purpose/aim of *personal development* and perhaps *preparation for dating.* Chemistry is not in this usage a language describing the impetus towards intimate interconnection involving hetero-sex or hetero-coupledom, nor can it be seen as merely a truncated version of this impetus. Understandings of chemistry as directed towards romantic/sexual hetero-intimacy are in this formulation at least deferred, perhaps thwarted and undoubtedly dethroned from its usual paramount position in most daters' stories. Instead, chemistry is conceived as foregrounding the singleton's experience of frisson in relation to others being productively directed towards the self.

Young daters may still experience excitement and the thrill of possibility (the chemistry associated with a romantic/sexual agenda) but may prefer to evade or delay contact to retain some control over the pace of intimate progression—as women daters often do—as well as learning about themselves and about how they might interact with possible partners in the future. This kind of personal development and preparatory aspect of internet dating is however not restricted to young daters. Many daters in blogs and other online media formats point out that internet dating offers a platform for gaining confidence, skills and even—in opposition to the view of the Communications academic quoted earlier—eventually improving one's chances of meeting a suitable partner, whether or not face-to-face dates are deferred.

> As an introvert, dating apps were a form of escape when it came to putting myself out there in hopes of meeting women. Not only was it easier to carry a conversation since I wasn't being subjected to face-to-face interaction, but it was also an opportunity to adopt a new, better 'version' of myself. I saw it as the chance to turn into the person I always wanted myself to be . . . I'm grateful for the time I spent on online dating. Without it, I would never have gotten opportunities to break out of my introverted

> shell. I also wouldn't have had the chance to develop my personal voice in social interactions.
>
> *(Jake, Australia, online magazine story, 2019)*

> Having multiple options to choose from will make you more confident to go after exactly what you want . . . As you mingle with multiple personalities, you will become better at dating. Getting a date is one thing; know how to carry yourself, the questions to ask to find out what you need to make a decision can be a different thing altogether.
>
> *(Robin, Australia, post on general blog, no date)*

Both Jake and Robin talk about self-development and increased confidence as benefits of time spent online.

Another element of this apparent disengagement from the heteronormative and heteronormative chemistry is that the practice of online dating is linked with a lessening social disdain towards the single life. As we noted earlier, the way in which chemistry is usually understood in internet dating tends to disparage singles as deficit insofar as it centralises the value of hetero-intimacy, and in the process disallows other possibilities. What we have called previously in this chapter 'the sting in the tail' lurking behind chemistry's joys for those who fail or refuse to comply with compulsory heterosexuality, is found in the constant pressure towards hetero-intimacy with its looming threatened punishments of failure, isolation and loneliness. However, such denigration of singles—of those who do not date or date without intending to move towards sexual or romantic intimacy—is rather offset by evidence that the social norm of hetero-intimacy is not necessarily the vital centre of internet dating for all users. Many daters, particularly young people, are apparently not bound by this focus. The social norm of hetero-intimacy is at least diminished by the sheer availability of multiple dating partners online. Internet dating ironically enhances the confidence to remain single by reducing the sense of being forced to hetero-partner. As Robin notes, compromise in partnering becomes more likely as the easy path, 'even if the current [person] is not entirely what you want', when 'you seem to have no choice' *(Robin, Australia, post on general blog, no date)*. Conversely, the greater the sense of choice, the less need to rush into partnering.

Hetero-sex and hetero-coupledom may reign in internet dating, as they do in all forms of (heterosexual) dating, but singledom—including singledom associated with evading or delaying face-to-face dates—becomes a more acceptable path if there are multiple choices available. The access to choices on internet dating undercuts notions of being 'stuck on the shelf' or 'past your use-by date'. The unprecedented capacity to meet up and partner that is basic to internet dating as a technology of intimacy oddly renders socially acceptable the practice of considering dating options but not necessarily actually going on dates, or going on dates but without intending to have sex or form a relationship. These deliberate and ongoing—even if not consciously socially purposeful or permanent—practices are in some senses subversive in that they unsettle or interrupt social norms regarding gendered heterosexuality. However, when we expand the meanings of chemistry in internet dating to include singles' experience of frisson without connection as a means to self-development, further possibilities beyond such modestly subversive practices are brought into view. The disturbance of heteronormative assumptions that arises in evasion and delaying practices shaped by purposes/aims other than hetero-intimacy, reveals other practices which question the very borders of the heteronorm and hence may be described as dissidence.

Dissidence, we suggested, can entail sustained and deliberate *alternative* (even though possibly temporary) directions which in some way challenge the boundaries of the (gendered) heteronorm, but which nevertheless do not necessarily reach so far as a refusal of gendered heterosexuality associated with heresy. In the Introduction (Chapter 1) we noted that subversion blurs into dissidence, but that the latter is more radical and clearly intentional in its departure from heteronormativity. Radical and intentional in the setting of this chapter does not inevitably involve conspicuous, sensational or activist forms of departure from orthodoxy. In the micro-politics of heterodoxical developments concerning intimacy what we see instead is evidence of alternative directions that do not simply postpone hetero-intimacy with a redirected focus on self-development that emphasises the excitements of self-affirmation and grooming of the self. Rather these largely unobtrusive dissident practices amount to a *replacement* of hetero-sex and hetero-relationships with alternative purposes/aims that are concerned with non-heterosexual interaction—that is, cross-sex fun and friendship. Many blogs and other sources substantiate the significance of these alternatives for internet daters, as the following examples illustrate.

> I use a ABSOLUTELY FREE site (sic) . . . has large numbers of Australian girls . . . lots of girls to chat with and quite a fun site (sic).
>
> *(Sandy, Australia, discussion forum post, 2007)*

> While I'm still yet to find a lasting romantic connection, I'll always be glad that I swiped right on so many of my 'failed' dates—I would have missed out on so much otherwise . . . in fact, my greatest successes from online dating have been the friends I've ended up making . . . when I started going on dates with women for the first time . . . the lines felt even more blurred, as someone who was just beginning to explore that side of my sexuality . . . I've been thrilled to connect with awesome women, some of whom have become my closest friends.
>
> *(Gina, Australia*—cited in Nguyen 2018*)*[16]

> Dating sites aren't all bad—a lot of people go on them to meet new friends.
>
> *(Martin, Australia, discussion forum post, 2007)*

As we have noted in regard to employing internet dating for self-development, it would seem highly condescending to describe such gently dissident practices as 'procrastination', or as inferior surrogates for hetero-intimacy. These practices clearly involve satisfying, relatively common, though normatively unconventional social options (Arndt 2015a, 2015b; RASA 2017; see also Roseneil 2007, 2004, and Chapter 3). Indeed, dating advisors, counsellors and 'coaches' are also frequently inclined to recommend that internet daters should try to enjoy themselves and not focus solely on meeting 'the one'. These professional commentaries sometimes just baldly state that 'many people meet new friends [online]' (RASA 2020) or, less often, even acknowledge that realistically most internet daters may not find a sexual/romantic partner online. It is common for professional commentators to offset these suggestions by reintroducing heteronormative outcomes, implying or asserting that finding the fun in online dating may improve one's chances of locating a partner. As an American 'development coach and communication expert' puts it: '[a]ll the success stories I've worked with are couples who had no expectations and just went in to have fun. They let it happen organically and didn't try too hard' (McElfresh 2020).

Their recommendations to have fun are often accompanied by assumptions of a requirement to partner (such as, daters will be more 'successful' if they just 'relax'), or reinforcing notions of heteronormative 'failure' (such that daters might have to accept 'second-best' fun, rather than partnering). Nevertheless, these recommendations do not always simply continue to promote heteronormative outcomes. Ironically, professional commentators may sometimes support non-normative practices, even if they seem unlikely purveyors of unorthodox directions. Heterodoxical practices and exhortations towards social change sometimes arise from highly unexpected places. The example below is, for instance, from a controversial Australian men's rights activist, online dating coach and media commentator who has been strongly criticised for promoting regressive views of gender and sexuality (Henriques-Gomes 2020; Sakkal 2020). Yet here, she offers support for dissident options. Though this advice to daters commending continuing to date for fun and friendship—despite not finding a sexual/romantic partnership—arises in an advice blog provided on a dating website, it also appears not to reiterate heteronormativity.

> According to RSVP's Date of the Nation research, just under one in three of singles have had short term relationships through online dating, with similar numbers having made new friends . . . You're much more likely to find a good friend who's happy to snuggle up and watch the latest episode of MasterChef with you, or go for the occasional bushwalk . . . That's the reality and it's still a great outcome; provided you rid yourself of the idea that you are on a desperate mission to find the perfect partner. It's far better to see the whole thing as a chance to expand your social circle, enjoy

> getting dressed up to go on dates, and have someone to come with you and experience pleasures together you'd be reluctant to do on your own . . . Enjoyment is the key—it's supposed to be fun.
>
> (Arndt 2015a)

However, at this point we become increasingly aware of the limitations of existing language to describe the dissident options which arise in internet dating. Sarah Ahmed (2006; see Chapter 2) has outlined a range of terms which signal the ways in which queer challenges to heteronormativity are suppressed. In a related fashion, Tulia Thompson explores terms which delineate less radical movements between normative and non-normative practices (Beasley et al. 2015: 690–692; see also Chapters 3 and 4). Their work is suggestive for our analysis of heterodoxical developments in heterosexual internet dating. In this context, we proffer some potential new ways of speaking about practices of intimacy that are largely invisible in mainstream popular and scholarly discourse, an invisibility which underscores the lack or erasure of a vocabulary to refer to meaningful relationships which do not conform to the hegemonic romantic/sexual model.

Daters and professional commentators draw attention to the largely unheralded emergence of intriguingly unorthodox directions regarding 'partners' in the relatively new technology of intimacy offered by internet dating. One of these may be described as a cross-sex friend and the other we might describe as a cross-sex platonic paramour, the possibly somewhat sexualised/eroticised but non-sexual social companion who provides an alternative model of partnering without sexual intimacy. Both of these options, friendship and platonic companionship—which may in practice be difficult to distinguish from one another—offer unconventional departures from hetero-sex and hetero-coupledom, undoing and redefining usual understandings of chemistry. The latter however refers to a more obviously radically dissident partnership that could invoke the term 'queerplatonic' (QP), in that it refers to a relationship which is not centrally organised around fun, friendship or sexual connection. It is more emotionally intense, significant, committed and intimate than is considered 'normal' for a relationship built on casual enjoyment or friendship, and provides ongoing companionable coupledom. Yet, it does not resemble the heteronormative standard for a romantic–sexual couple (Urban Dictionary no date; Smith 2018). All the same, we are more inclined to describe this non-normative coupledom as 'alternative-platonic' (alt-platonic) or 'heterodox-platonic' (AP or HP), since the language of 'queer' most often denotes an 'opposition to hegemonic norms', a protest against the 'normal' (Showden 2012: 8; see also Beasley et al. 2015; Green 2007: 28–29). Such non-normative modes of cross-sex coupledom are rarely constituted as consciously socially oppositional. For example, Suzie, a UK-based counsellor, realised in her late 50s that she was craving companionship and affection but had lost any interest in having a sexual partnership. This is not expressed as a rejection of her previous heterosexual relationships.

> I'd been married and had great sexual relationships, but I got to the stage where other things became more important . . . It was a very natural process—it wasn't a conscious, 'Oh, I'm not going to have sex again.' It just happened . . . I moved house, came out of a relationship, changed career . . . and, when several years had gone by, I realised I was content without it.
>
> *(Suzie, UK*—cited in Knight 2013*)*

The development of cross-sex friendships or AP/HP/QP relationships can be particularly important to older daters: 'One of my friends met several older gentlemen through online dating sites after her husband died. She had platonic relationships with a couple of them, eventually become cruise companions! (sic) . . . she enjoyed the company' *(Beatrice, UK, advice article on social networking site for over 50s, no date)*.

These cross-sex relationships are by no means inventions of the internet and have an extensive if under-researched history. Examples may be found in long-standing cross-sex partnerships to undertake hobbies such as going to the cinema and playing card games like bridge or sports like tennis (see for example, Brkljačić et al. 2017). Nonetheless, we suggest that internet dating, because of the unparalleled access to dating options it offers, may not only reduce the fear of singledom as pointed out earlier but additionally provide unprecedented opportunities to take up innovative forms of intimate social connection with relative ease. These opportunities are simply less available in offline dating. Internet dating etiquette of several dates before deciding to have sex and of multiple parallel dates provides fertile ground for an awareness of affectionate non-sexual chemistry that may generate alt-platonic (AP) coupledom. Dissident innovations of this kind may not so much challenge or defy heteronormativity as displace it with un-like alternatives—typically more everyday, even mundane, rather than socially provocative ones. This has implications for more nuanced characterisations of dissidence and social change *per se*, reminding us that innovations in the micro-politics of intimacy are not necessarily heroic, scandalous or momentous. In this setting, the dissident innovations outlined here yield 'both/and' rather than 'either/or' departures from the norm (Elbow 1993: 22–24) and bring to life what Michael White calls 'subjugated stories' (Carey et al. 2009; White 2005). We have come a long way from well-established notions of chemistry as knowing immediately that your potential date is your sexual/romantic 'soul mate'.

Conclusion

In this chapter, we focussed our analysis of possibilities for delimiting gendered heteronormativity and advancing gender equality in internet dating upon what amounts to the quintessence of seeking intimacy online—chemistry. While without doubt chemistry is viewed by daters as *the* vital attribute of their engagement with internet dating, characterising this hallmark proves to be decidedly elusive (see also Finkel et al. 2012).[17] The daters in our sample seem convinced it is the crowning glory of their endeavours, the fundament of their motivation and their goal, and yet it appears as a portmanteau term including a bundle of possible features and much more slippery than the certainty of its significance in daters' narratives would seem to suggest. An added difficulty is that, since we are centrally concerned with innovations in intimacy, we must attend to both common and uncommon understandings of it. In summary, we have characterised chemistry as about beginnings and referring to a mesmerising sense of immediate and special, even unique affinity. Though this most often describes sexual connection, chemistry is not necessarily sexual or even mutual, let alone leading to a lasting outcome. And, in the setting of internet dating, chemistry may begin online but is almost always reliant on embodied affirmation offline.

It is apparently ineffable and evidently an almost inarticulate, less than conscious, embodied experience. Perhaps this goes some way to explain why the notion of chemistry has such a powerful hold over daters' imaginations, dreams and wishes. Yet, by contrast, this crucial feature of internet dating is under-researched and under-theorised. Because of its significance in the lives of daters and conversely its comparatively thin level of scholarly attention, here we bring chemistry into the foreground. Given our core concern with social innovation, we considered the ways in which this notion can reiterate and sustain gender norms and heteronormativity more generally, as well as generating directions which challenge gendered hetero-sex and hetero-coupledom. Chemistry, it seems, can ignite a thrilling frisson, flirtation, passion and love but also, perhaps rather surprisingly, heterodoxical practices ranging from subversion to dissidence that can even challenge the very basis of internet dating.

Notes

1 Finkel et al. (2012: 3–4, 6–7) pursue an evaluation of the possible differences between and the benefits of internet dating over offline dating within a relatively strict rubric which focuses largely on technical issues: they evaluate three issues—access, communication and matching. While their study is attentive to the concerns of internet daters, it is primarily attuned to utility rather than any socio-political analysis.

2 Like Miller et al. (2016: xiv–xvi, 21), we remain unconvinced by approaches such as that of Rainie and Wellman (2012) which assume that internet technologies inherently support individualisation. We note, in this context, Miller et al.'s account of both individual innovation against the grain of gender norms along with substantial use of social media to uphold such norms in their comparative investigation of nine different cultural sites in eight countries.

3 Similarly, in the wake of the #BlackLivesMatter movement, some dating sites/apps have undertaken technological renovations on the basis of advancing social

equality in the formation of intimate relationships. *Grindr* and *Bumble* among other sites/apps have reconsidered their so-called 'ethnicity filters' to prevent daters stating ethnic/racial preferences which then filter out those who do not fit these preferences. By contrast, the world's largest internet dating company Match Group Inc. defended such filters on some of its many brands (such as *Hinge*) as actually of benefit to minority communities and as assisting them in resisting mainstream racism—even though another of its platforms (*Tinder*) has no such filter (Petter 2018; Stokel-Walker 2018; Zelaya 2020). This development, like *Tinder*'s 2016 update of gender terminologies, underscores that the technology of dating platforms is not divorced from the meaning and practice of chemistry.

4 We stress however that 'chemistry' is sometimes a term used to describe a couple's sense of continuing magnetism. For example, one Australian blog 'addict' uses the word in both the sense of immediate beginnings and ongoing compatibility. In response to a negative assessment of internet dating, he insists that it is worthwhile and points out that he had a 'lucky break' meeting his wife on one of the mainstream dating sites: 'I was pleasantly surprised, I had a few dates with OK gals, no weirdo's, just no chemistry. With my now wife, the chemistry and attraction was there straight away (and still is obviously) [sic]' *(Shane, commenting on Eric's post, Australia, general blog, 2007)*. We emphasise the temporality of chemistry in the context of internet dating simply because in daters' accounts it is typically a way of describing (often noted with some amazement and a feeling of unexpected good fortune) the sense of the immediacy of connection at an early stage. (Our thanks to John Gray for discussions on this point.)

5 It appears that the respondents were not asked about whether their sexual partners were same or different sex.

6 The so-called 'pro-sex' stance within feminist, gender and sexuality studies specifically resists a risk-oriented approach to sexuality/sexual relationships and embraces pleasure and sexual exploration (Beasley et al. 2012). For more discussion of the 'pro-sex' approach see Chapter 4 and Chapter 4, note 3.

7 We wish to acknowledge here conversations with and the work of Andreas Henriksson. In a 2017 publication, Henriksson and colleagues have drawn attention to the ways in which the social activities of older people are reduced to their utility and are thus diminished and trivialised. We have reworked this approach somewhat to discuss internet dating (see Krekula et al. 2017).

8 For a more extensive discussion of the less appealing side of social narratives around chemistry, see Beasley and Brook 2019: 146–148, 154–155.

9 Income appears to be of increasingly lesser importance for both women and men, at least in studies of 'Western' internet daters (Dinh et al. 2018: 9). However, this development is not necessarily relevant for other cultural contexts such as China where wealth and associated status is still highly pertinent for both men and women (Li and Lipscomb 2017).

10 Many, if not most studies of internet dating forcefully and uncritically reiterate gendered heterosexual binaries such as aligning men with pursuit of sex and women with love, as well as aligning men with preferring young women and women older men (see Bruch and Newman 2018). Yet, it is worth noting that McWilliams and Barrett (2014), who unusually examined gendered power relations, found that both men and women declared that they would prefer younger partners, if for different reasons. We wish to offer another perspective in which daters' asserted views may be more gender normative than their practices.

11 For an analysis of age-dissimilar relationships see McKenzie 2015.

12 Data on gendered age differences between heterosexual couples in the USA and the UK is similar to these Australian statistics. The global average of such age

differences is around four years, which includes countries with much larger age gaps (Pew Research Centre 2019).

13 Less than positive terms which are used to describe heterosexual partnerships between a much older man and younger woman suggest a degree of discomfort with or even disdain for relationships which are deemed to involve too great a discrepancy in wealth, experience, status and power and thus at odds with the Western ideal of mutuality in love matches. The terms include cradle-snatcher, lech, sugar daddy, gold-digger and trophy wife.

14 Women in this American study were five times more likely than men to consider that they were sent too many messages (Vogels 2020).

15 The capacity to prolong the online stage of internet dating is precisely noted as one of the significant benefits of dating online for older adults, especially women, in an assessment by Viveros and Schramm (2018).

16 Gina is not the pseudonym used by the author of this reference. We have provided an alternative one to avoid confusion with another dater cited previously.

17 Finkel et al. stress that research on intimacy is still emerging, but they also appear to suggest that understanding of the development of relationships has advanced even if predicting romantic outcomes remains obtuse (2012: 3, 39). We are perhaps less convinced that most research on internet dating has been productive of enhanced understanding concerning personal relationships.

Chapter 6

Older internet dating: Over 50 and beyond

Although internet dating is a relatively new phenomenon making use of comparatively new technological features which affect access to potential partners, as we observed in the opening remarks of the Introduction (Chapter 1) the mode of instigating connections and matchmaking formats (Finkel et al. 2012: 6–7), is not entirely dissimilar to previous forms of matchmaking in many ways. Yet, when we turn our focus to older daters, we see some novel developments and possibilities that exceed technological innovations and contribute to our discussion of intimacy, online dating and social change.[1]

Internet dating is an increasingly popular means to developing intimate, usually sexual relationships, but importantly it is increasingly aimed at assisting in the development of such relationships for *all* age groups (Barraket and Henry-Waring 2008: 150; Davis and Fingerman 2016). This shift has been accompanied by a burgeoning of 50+ dating sites. These designated sites in Australia and the UK—such as *SeniorPeopleMeet*, *Singles50*, OurTime, *Over50s*, *SilverSingles*, *Lumen* and *MatureLove*—are variously described as focussed upon 'older adults', 'middle years' and 'later years' or as 'mature' and 'senior' dating, and have arisen alongside specifically targeted sections within general dating sites (like *eHarmony*).[2] Relatedly, an ever-growing number of older people are making use of online dating services, with divorced older singles being considerably more likely to use them than other older singles (Viveros and Schramm 2018). Indeed, in the UK, Australia and many other countries, the fastest growing sectors in the internet dating market now include younger people (18–24) and older adults—characterised as those at least over 50. Between 2013–2015, according to a Pew Research Centre study, the number of users aged 55–64 doubled (Shaw 2020; see also Elin 2014; Bowling Green State University 2012).

This expansion of the online matchmaking market coincides with unprecedented changes in the social demographic across the globe. There is an ever-growing cohort of older people in the world—including *single* older adults—with an expanding life expectancy. More than at any other time in human history, older singles are meeting up to form new intimate/sexual relationships. They are now very much the subjects of matchmaking services and choosing to avail themselves of these services (Davis and Fingerman 2016). Not surprisingly, in *The Modern Dating Report* (The Australian Seniors Series 2018)—an ongoing national study which examines the views of Australians over 50—eight out of ten of these older adults considered their generation were more likely to date than in the past.

It is still the case that the great majority of those undertaking online dating are under the age of 30 (Paisley 2018).[3] In 2020, 85 per cent of users worldwide were under 34 (Dolan 2020). Younger internet daters are also more likely than older ones to state that their marriage or committed relationship began online (Smith and Duggan 2013: 5). However, the proportion of older single users is rising. The numbers of older people that are making use of the internet is increasing in many countries. In the UK, the gap between them and younger age groups is closing with the 65 to 74 age group showing an increase in recent internet use from 52 per cent in 2011 to 83 per cent in 2019. Older Australians are also increasingly online (Australian Bureau of Statistics 2018; Office for National Statistics 2019a). There appears to be worldwide increased use of internet dating as the means for older people to undertake intimate/sexual relationships (Malta, cited in Cooper 2008; see also Malta 2013).

These shifts in the practices of older people—primarily older singles—are highly relevant to our concerns in this book. The book overall is directed towards examining heterosexual accounts of dating in order to explore signs of social change in 'the realm of the dominant' and to this end has so far investigated various aspects of online dating (nutters, norms, excitement/emotion, chemistry) in previous chapters to bring to light heterodoxical directions that in some ways depart from the hegemony of the gendered heteronorm. However, while we retain our overarching concern with the micro-politics of social change in the domain of hegemonic heteronormativity, in this chapter we alter our vantage point to gain another means to consider the heterodoxical. We turn now to examination of a subgroup within internet dating that occupies a marginalised rather than mainstream position within heterosexual internet dating. Older heterosexual daters are comparatively rarely the subject of scholarly attention in research on internet dating as this literature almost invariably concentrates upon and privileges the young (Huyck 2020; McWilliams and Barrett 2012: 412). In this chapter, we deliberately depart from following the predominant research focus on young daters. All too often when discussing dating, sex and love, the images conjured up by these words involve young people starting out on adulthood, smooth young bodies, sex and sexiness as necessarily an attribute of youth, and the discoveries of the young as they face what for them are new experiences. These everyday unremarkable images, which seem so self-evident, harmless and charming, nevertheless exemplify assumptions about what constitutes intimacy and intimate relationship, thereby reproducing systemic prejudice against older people that discounts their complexity as human beings, as irreducibly social subjects, as relational and interconnected bodies, and as socio-political agents. In this chapter, we focus on older heterosexuals precisely to extend and give further substance to our preoccupation with opportunities for change in relation to gendered heteronormativity.

As was initially outlined in Chapter 1 (Introduction), while heterosexuality is hegemonic, relations of dominance do not describe heterosexuality in its entirety. Rather than viewing heterosexuality—with its accoutrement of binary hetero-genders—as an undifferentiated monolith, in the previous chapters we have explored varying aspects of online dating which reveal its heterodoxical possibilities. In keeping with this approach, we concentrate upon older heterosexual daters specifically to foreground *variability within an arena of social dominance*. Relations of dominance are not all of a piece; heteronormative hegemony is no different. Its patchy irregularities give weight to our view of gendered heterosexuality as a diverse range of practices that include heterodoxical departures. These departures provide occasions for change and advancing social equality, including shifts from gender normativity towards enhanced gender equality. Focusing upon older heterosexual daters enables attention to them as a marginalised cohort that might provide intriguing and less often explored instances of fissures in the realm of the dominant.

Definitions and broader characterisations of 'older' vary significantly within scholarship, popular culture and dating sites/apps (Field 2018: 2; Tarrant 2010).[4] For us, this variability indicates that the cohort is not strictly about biological years and supports the perspective offered by King and Calasanti (2013: 699) that age is also associated with contextual social assessments such that the conception of older age has 'a political location'. Because of the socio-political meanings attached to naming this cohort, we have elected to employ the most inclusive use of the terminology and seemingly the least associated with pejorative connotations, referring to all those over 50 years as 'older' as this is often the turning point employed by dating sites and apps. Researchers in the field are sometimes inclined to offer a more nuanced assessment of those in their 50s and even early 60s as 'middle' or 'mature' aged and many acknowledge that the experience of the so-called 'young old' (50s–70s) might differ from those who may described as 'elderly', or 'old old' (75+; Pinsker 2020). Nevertheless, we make use of the pragmatically inclusive vocabulary of 'over 50' to encompass a broad cohort who are distinguished in market, social and cultural terms as 'older' and are registered as such by internet dating services. By this means we aim to shift the usual attention given to young daters as somehow exemplifying what dating is about and instead bring into view the particular experiences of older daters, and more specifically older singles.

In this context, it is important to note that the growing development of dating services for and use of internet dating amongst older people is not a matter of esoteric interest but rather of increasing social significance, given that by 2051 it is predicted that 25 per cent of the Australian population will be aged over 65 (Malta 2007: 85) and similarly for the UK (MacInnes and Pérez Díaz 2009). Moreover, according to the World Health Organisation (WHO) the 'greying' of the world's population is global (WHO 2002: 6–7; Uhlenberg

2009a: 1–2) and the growing number of older people—including the rising number of single older adults—will be accompanied by a parallel increase in life expectancy (Malta 2013). Population ageing is accelerating and involves increasing numbers of people living for many years after present retirement age. For example, in Australia 'a female (male) born in 2012 could expect to live for 31 (29) more years after they reach 65 years old' (Australian Productivity Commission 2013: 33). In the UK from 2016 to 2018, 65-year-old men could expect to live another 18.6 years and 65-year-old women another 21 years (Office for National Statistics, 2019b). These older people will not necessarily be dependent and may well continue to live active lives (Spijker and MacInnes, 2013). Yet, despite the growing significance of internet dating in relation to older people (Hogan et al. 2011), literature on this topic is limited (Field 2018; Menkin et al. 2015). Popular, advice-oriented and scholarly literatures on ageing *per se* and those studies which are available on older online daters tend to highlight health risks. Health risk concerns such as those associated with ageing populations and increasing dependency are, for example, prevalent throughout *The International Handbook of Population Aging* (Uhlenberg 2009b; see also Commonwealth of Australia 2019 and UK Parliament 2015) and are also notable in studies and reports on older-aged online daters (Alterovitz and Mendelsohn 2009; Bateson et al. 2012; Brown and Shinohara 2013; Family Planning NSW, no date; Griffiths 2018; Noone 2012; Shaevitz 2015). This health-centric/medicalised characteristic of analyses on older dating is especially evident when considering 'desirability', bodily intimacy and sexuality—a point to which we will return (see Gewirtz-Meydan et al. 2018). Indeed, as noted in Chapter 5, attention to utilitarian health concerns in analyses of older people and internet dating is such that older people are frequently constituted as a health problem associated with deficit, loss and decline. This diminution has significant implications with regard to provision of health and care services cognisant of older people's intimate connections with others, but also seeps into understandings of their use of internet dating.

While in recent years more attention has been given to the social contributions which older people have made and continue to offer (Gong and Kendig 2016: 24–25), a broadly deficit model orientation remains in evidence in examinations of older people's lives and relations with others, particularly with regard to intimacy/sexuality, which has implications for understandings of the conduct and meanings of their uptake of internet dating. Hence, we initially outline the general contours of this deficit model and then begin to examine whether internet dating offers opportunities for heterodoxical directions exceeding gendered heteronormativity. Notwithstanding evidence of ongoing norms which continue to constrain how older people undertake and experience internet dating, there is more to be said here. In this setting, we then outline normalising parameters of ageism with regard to technology, intimacy/sexuality,[5] health-centric and medicalised narratives and notions of inflexibility and social quietism. In considering these restrictive parameters, our intention is to explore how older people might creatively escape heteronormative conventions and thus to link their intimate/sexual practices in internet dating to social change.

The 'Western' deficit model of older age: Ageism

Ageism takes the form of dismissive and stereotypical attitudes and beliefs, discriminatory behaviours and formal policies and practices which overestimate negative elements of older age associated with 'poor health, financial concerns and dependency' and simultaneously underestimate the contributions and positive features of older people's lives (Australian Productivity Commission 2005: xii; Wallace et al. 2017: 9). Indeed, despite some emerging evidence of more constructive perspectives, 'the view of old age as a defeated stage of life'—which Kendig and McCallum considered would be overtaken by social change when the baby boomers 'joined the ranks of the aged' (1986: 59)—has by no means disappeared. While ageism is not a given across the globe and varies by social context, it is found in both developing and developed countries and cultural traditions of respect for elders do not mean ageism is absent (Chang et al. 2020; EAC no date; WHO no date; Vauclair et al. 2017). There is a significant body of ongoing research which stresses that 'ageism is pervasive, widely accepted, and normative in many cultures and societies . . . and is a form of prejudice that goes unchallenged, and even celebrated in many fields' (Sargent-Cox 2017: 5).

Importantly for our purposes ageism is strongly associated with significantly worse health outcomes along with other pernicious social impacts and inequities (Chang et al. 2020; Roscigno et al. 2007: 313–314). Declining social status of various kinds is a feature of growing older in the UK and Australia, associated with a 'problematisation' of older adults as being out of touch, frail, a burden and a cost (Officer et al. 2016: 710). However, such problematising is also inclined to encourage the development of a self-fulfilling prophecy: research studies concerned with older people have found that they often 'reflected an internalization and acceptance of ageist stereotypes and prejudices', such that being old was linked with a string of unappealing attributes including: 'not trying, withdrawn, isolated, irritating, self-oriented, living outside the mainstream, unattractive, uninteresting, frail, senile, silly, over the hill, narrow-minded, a burden, lonely, vulnerable, dowdy and unproductive' (Minichiello et al. 2000: 259).

In short, the impact of age discrimination upon the health and status of older people, as well as upon prejudicial perspectives held about and by older people, has implications for their engagement in dating *per se*. It is impossible to examine the potential for social innovation and opportunities for advancing gender equality amongst older internet daters without attending to the prescriptive normative pressures they face associated with ageism.

The ageist deficit model in 'Western' societies like the UK and Australia has impacts upon all that fall within its rubric, even those for whom gender or other axes of power have formerly attached privilege (King and Calasanti 2013: 699–700). In this setting, the social status and experiences of heterosexual men and women may converge somewhat as they age, although ageism is gendered, with men, for instance, suffering a loss of work-related status and women losing status related to youthful attractiveness but gaining advantages from the social networks and related emotional intimacies they have maintained (Beasley and Holmes 2016: 213–214). Older heterosexual men and women both find themselves shifting in their social status—especially with regard to hetero-gendered notions of embodied capacity, attractiveness and intimate/sexual desirability—and hence to a degree sharing rather more ground than they might have in their youth (Hughes 2011; Bennett 2007; Hearn 2007; Rhohlinger 2002). The question is, do these shifts in social positioning associated with ageing lead to heterodox forms of gendered heterosexual intimacy? Can the novel terrain of internet dating provide a space which enables a reassessment of the deficit model and in which potential non-normative innovations might arise and even flourish?

Ageism and technological deficit? Intimations of change

It may be of course that internet dating is not necessarily a site for innovative possibilities for older daters. There are signs that what Sara Ahmed (2004: 11, 60, 89–95) describes as 'sticky' affects—that is, intransigent attachments to 'traditional' social interactions, in this case traditional conceptions of romantic 'meeting up'—may still be in play (Whitty 2008a; see also Chapter 2). Many women internet daters find themselves passively awaiting messages (Frohlick and Migliardi 2011: 79–81; Moss 2013; Smaill 2004: 94–96) and this may especially be an issue for older daters. Claire, for example, a 53-year-old participant in a study by McWilliams and Barrett (2012: 422), acknowledges that she too is an 'awaiter of messages' (Moss 2013) and only actively approaches potential dates as an expedient last resort: 'I almost never contacted people. When I would be bored and deciding nothing was happening, occasionally I would contact people' (McWilliams and Barrett 2012: 422). Moreover, a range of studies indicate that older heterosexual adults face gendered cultural mores associating attractiveness with youth and thus older women felt they were required to employ a range of cosmetic enhancements designed to generate online attention. This concern with appearance on the part of older women is seemingly not misplaced since older men commonly place considerable emphasis upon women's looks and youthful femininity (Clarke and Griffin 2008; Field 2018; Gewirtz-Meydan and Ayalon 2018; McWilliams and Barrett 2012: 416, 425).

Internet dating reiterates gender normative mores in a range of ways but this reiteration is strengthened by assumptions that older people are technophobic 'digital immigrants' and have little knowledge, skill or interest in new technologies (Adler 1996; Cooper 2000; Fong et al. 2001; Philbeck 1997; Prensky 2001). Older people are often perceived as inflexible, narrow in their thinking and as incapable, inactive and/or uninterested in a range of arenas including those associated with desire and love, and hence are not viewed as being intrigued by new connections with others or by romantic/sexual relationships, but before we turn to these constricting aspects of ageism it is important to recognise that such perceptions have implications for how older people are seen as engaging with dating technologies.

As Malta notes, computers have long been conceptualised as 'the exclusive bastion of the young' (Malta 2007: 86) and similarly the internet itself and all internet device usage remain markedly associated with youthful creativity, open-mindedness and exuberant sociality. For the most part, older adults have been seen as struggling to keep up with technological changes or often as having been 'left behind' (see Furlong 1989: 145; White et al. 1999: 359; Fong et al. 2001: 3). More recently, such an account has been subject to question as the internet has become ever more ingrained in the lives of people around the world. It would seem that the number of older people accessing the internet is steadily rising and once the shift to online technology is made many of these older users become as engaged as younger people. Older users of the internet have been particularly inclined to take up email, as well as online shopping, entertainment, banking, health and government information and advice, education, travel planning and a range of other activities such as joining online communities. Older people have so far been rather less inclined than younger users to publish and share online content, particularly audiovisual content, and their adoption of internet-based activities is strongly affected in the UK and Australia by income among other issues (Milligan and Passey 2011: 29; Ketchell 2018). A 'digital divide' continues to exist. Those in higher socio-economic groupings are 'more likely to use the internet than those in lower groups' (AgeUK 2016). Yet this divide reflects several axes of social inequality rather than being intrinsically or solely a matter of age. Indeed, uptake of online activities by older people is growing at a rate that is scarcely indicative of an unmoving technophobic refusal or mental obstinacy.

Though it would seem that older people are somewhat selective in their adoption of expanding new technologies, the marked paucity of research on older online daters as against on young users (Field 2018; Malta 2007; McWilliams and Barrett 2012) cannot be justified on the basis that those over 50 are unresponsive to such technologies. In the UK, 52 per cent of the 65 to 74 age group reported recent internet use in 2011; by 2019 that had risen to 83 per cent. In 2019, 47 per cent of over 75-year-olds were recent internet users, compared to 20 per cent in 2011 (Office for National Statistics 2019a). In Australia in 2014, 68 per cent of those over 65 had accessed the internet in the last six months and by 2016–17, 55 per cent of this age group had accessed the internet in the last three months (Australian Bureau of Statistics 2018; Australian Communications and Media Authority 2018). And, as mentioned at the beginning of this chapter, older people are evidently seeking to expand their social options though undertaking internet dating in ever greater numbers. While older users frequently point out problems they experience in internet dating (Viveros and Schramm 2018; Vandeweerd et al. 2016), they also see it as expanding their social options in the face of diminishing opportunities to encounter potential partners. For 62-year-old Lillian it provides a crucial, incommensurable and even exciting means to be able to develop relationships with new people: 'You come home from work, you want to be talking to someone about your day, it's good for your mental health . . . For women of my generation you can't just walk into a pub by yourself' *(Lillian, UK*—cited in Shaw 2020*)*.

Many such users convey their pleasure in engaging with what they deem an accessible and effective technology of intimacy. As one older dater puts it,

> I had virtually no luck offline. First, I have no idea how to meet men [because] I'd been married for a long time. Second, I'm not gorgeous, and hate being rejected out of hand. Using [this dating site] was amazingly easy.
>
> *(Irene, Australia*—cited in Malta 2007: 94*)*

While those over 50-year-olds who undertake online dating are typically more familiar with using the internet than their peers, the presumption that older people are necessarily technophobes seems especially misplaced in the face of such comments. In the context of assumptions that using the internet, and especially using it for dating, is for the young, older internet daters are undertaking activities which challenge age normativity. Our concern here is that although older people are becoming more adept at and increasingly more involved in internet dating, their activities continue to be described almost entirely by the language of deficit, delay or obstinacy rather than as innovative in the face of prescriptive and often demeaning social norms.

Furthermore, older women daters may be viewed as departing from the limitations of *gendered* age normativity in that they are particularly considered to be incapable of engaging with, let alone effectively interacting with technologies. As Ulf Mellström has pointed out, 'technologies are to be understood as means of an embodied communication for forming [male] homosocial bonds. These masculine practices continuously exclude women and perpetuate highly genderized societal spheres' (2004: 368).

He notes that 'because technology is pervasively a masculine cultural expression' (Mellström 2004: 369), women's skills are typically not conceived as technical skills. Given this highly durable and ubiquitous equation between technology and masculinity (Faulkner 2000) and, despite some variations, the continuing significant gender gap in the IT industry and internet usage worldwide (Chakravorti 2017; Mellström 2009), older women's increasing engagement with negotiating the technology of internet dating in the UK and Australia, alongside their male peers, may be understood as a form of heterodoxical divergence, even of transgression. Innovations in terms of age and gender power relations to be found in over 50-year-olds' online dating require acknowledgement in order to avoid reiteration of ageist assumptions not only about their activities on the internet but importantly also in relation to older people's contributions to social change more broadly. This chapter is a contribution towards that acknowledgement, shining light on heterodoxies in older internet dating.

Ageism and desexualisation: Heterodoxical quandaries

By attending to heterodoxies associated with older internet dating we challenge the starkest sign of ageist evasion—that is, neglect of older adults' bodily/sexual intimacies. This neglect is most strongly registered in both popular and scholarly literature in a general diminution of older people through *desexualisation.* Desexualisation dehumanises but also prescriptively restricts heterodoxical possibilities and recognition of these possibilities, as can be seen in the limited attention scholarly research pays to older people's online dating *per se*—noted previously in this and other chapters. The propensity of existing research is to adopt a perspective which gives little consideration to older adults' exploration of intimate connections with others as well as the opportunities for social innovation that might occur in this exploration. The prevalence of the desexualisation of older people in this research amounts to a normative discomfort with and/or disinterest in older people's emotions and bodies concerning intimate relationships. Ageism produces a focus in internet dating studies on the young (see for instance, Clark 1998; Hillier and Harrison 2007; Hillier et al. 2012) and a restricted range of scholarly works examining how older people use the internet to develop intimate connections with others (such as Adams et al. 2003; Malta 2007; Malta and Farquharson 2014a). Studies that do exist on older online daters typically say little about the possible range of intimate connections that may be sought or found (though see notable exceptions like Debrot et al. 2017), or about desire (barring a notion of 'preferences' and 'attractiveness'), let alone directly about bodily intimacy, sexuality and sexual practices.

The focus of the scholarly research when it does deal with sex is often focussed upon deficits and associated health/safety concerns such as compensating for loneliness, exposure to unwanted sexual content or, most often, increasing vulnerability to sexually transmitted diseases—despite considerable evidence of the latter being a very limited health concern (see for example, Albury et al. 2020; Bateson et al. 2012; McWilliams and Barrett 2012; Noone 2012; Viveros and Schramm 2018; see also National Institute on Aging, US Department of Health and Human Services 2017).[6] As Sandberg observed in 2013,

> [i]t is striking how little qualitative knowledge exists on how older men and women experience sexuality while ageing. The voices of older people themselves are rarely heard and the existing bulk of knowledge is predominantly based on large scale quantitative studies conducted by geriatric and medical specialists, leaving little room for older people's own ways of making sense of sexuality.
>
> (264)

The scholarly literature largely continues to offer a diminished view of the interactive social meanings attached to how older people engage in online dating. It does this partly because older people are predominantly deemed not sexual, let alone variable in their sexualities, and consequently sexual expression amongst older people is frequently cast as problematic, strange or unseemly (Barrett et al. 2008; Brown 2009). Denmark (2002:17) notes that 'one of the most pervasive myths in our society is the belief that a decrease in sexual interest and a diminished capacity for sexual behaviour are an intrinsic part of the aging process' (see also Adams et al. 2003: 405; Malta 2007: 84). The ageism in much research means it cannot provide much insight into the everyday phenomenological ways in which older adults may undertake and experience intimacy/sexuality, let alone may challenge social norms associated with heterosexual relationships (Fileborn et al. 2017: 2097).

These ageist distortions leave a gap in existing scholarship around what McWilliams and Barrett (2014: 430) describe as 'micro-level interactions in the online dating environment' and we have termed the 'micro-politics' of heterosexual internet dating (see especially chapters 1 and 7). We foreground intimacy/sexuality precisely because these aspects of older internet dating are very frequently diminished, or are simply absent, despite their importance.

As we briefly outline in chapter 1, intimacy is a term usually employed to refer to a more diffuse arena than sexuality, evoking a sense of close, embodied and particularised personal connection (Budgeon 2008; Debrot et al. 2017; Debrot et al. 2013; Henriksson 2014; Jamieson 1999; Roseneil and Budgeon 2004), and involving individuals but also relationalities (Smart 2007). The term sexuality, by comparison, encompasses not merely sexual activities and experiences but also a conglomerate of social institutions and arrangements as well as subjectivities and social conventions (Beasley et al. 2012; Jackson 1999). In attending to the micro-politics of internet dating we conceive sexuality as located in the continuum of intimacy, though not as reducible to it. Here we focus our consideration of the connections between intimacy and sexuality found in embodiment and physical expression. Looking at heterosexual internet dating through the lens of *bodily* intimacy/sexuality provides a means to concentrate our thinking about older people's connections to others and the potential in those connections for contributions to gender equality and social change. Apart from bodily intimacy/sexuality providing a key to analysing the normalising parameters of ageism and its impact upon older internet daters, our focus upon physical connection also works against the tendency of such parameters to evade the ageing body—a tendency which maintains the centrality of youth's supposed equivalence with desire, romance and love. It is important not just to pay attention to physical connection between older adults but additionally to engage with the significance of and nuances in physical connection in order to tease out the micro-politics of potential social innovations.

Sexuality and sexual practices matter in discussions of older people and internet dating, as was mentioned in chapter 5, because there is strongly established link between sexual health and broader aspects of well-being. A failure to include sexuality in discussions of sociality has some highly problematic consequences given that sexual expression appears to be 'an essential aspect of our lives' (Barrett 2011: 32; see also Planned Parenthood Federation of America 2003; Seymour and Lupton 2004) predictive of a heightened quality of life (Weeks 2002). Yet despite the 'significant body of evidence linking sexual health to emotional well-being of older people', as Barrett (2011: 31) notes, 'few health or human service organisations have programs to promote the sexual health of older clients'. Internet dating may step into this gap.

Internet dating can facilitate intimate connections between older people. These connections matter to them, as well as enhancing their health, and such connections include but are not solely described under the rubric of sexuality. Our approach here is shaped by the work of Debrot et al. (2017) and Menkin et al. (2015) (see also Alterovitz and Mendelsohn 2013; Debrot et al. 2013) whose research illustrates that the health benefits to older people accruing to them from sexuality and sex cannot be disentangled from affection, such that companionable physical signs of intimacy mediate the significance of sex, and that across all age cohorts companionate intimacy was as, if not more, valued in relationships than sexual attraction, with this valuation increasing somewhat with age.

To be able to consider the full range of ways in which internet dating involving over 50-year-olds may reiterate or challenge gendered heteronormativity it is crucial to grasp how older adults negotiate ageist normalising parameters concerning desexualisation in a much broader sense than is usually associated with the term. Ageist desexualisation of older people does not only mean presuming older people are asexual but rather also conceiving older adults as normatively distanced or even disconnected from physical intimacy—as not having intimate bodies, let alone bodies inviting or seeking intimate touch. New approaches and terminologies are necessary if we are to capture heterodox possibilities in older people's lives. Consequently, we are inclined to make use of new vocabularies, such as the language of 'social flesh' we have employed previously (see chapters 4 and 5), which resists desexualisation and disconnection from bodily intimacy. Social flesh instead offers a framework for understanding sociality which is robustly attentive to bodies, intimacy and sexuality, while also enabling consideration of emotional reflexivity (see chapter 4; Holmes, 2010). The central point here is that sexuality *and* intimacy must be included in our deliberations concerning the social flesh of internet dating and social change.

Ageism and bodily intimacy: heterodoxical challenges

When we bring sexuality into view and pay attention to the overlap between sexuality and intimacy in investigating older people and internet dating, some possible signs in favour of heterodox innovations emerge. While few 50+ online daters might venture toward the more heretical, or 'queer' end of heterodox innovations, age may interact with new technologies to enable more possibilities for innovations that diverge from, transgress or subvert norms around ageing, intimacy/sexuality, gender subjectivities and heteronormativity. These innovations might be more easily found amongst the young-old (50–74) than the old-old (75 and older). For instance, one analysis of heterosexuals' online personal advertisements found the young-old more likely to mention adventure, romance, sexual interests, and seeking a soul mate and less likely to mention health (Alterovitza and Mendelsohn 2013). Similarly, comments in response to a posting about older women and internet dating on a discussion forum are in keeping with a number of studies in suggesting that older male daters are only interested in 'some good looking older women that visits [the] gym (sic)' (*Ivan, UK, discussion forum, no date*). Sexual activity amongst older adults is thought to require a youthful appearance (see also Featherstone and Hepworth 1991; Gewirtz-Meydan and Ayalon 2017; McWilliams and Barrett 2012). Nonetheless, there are other instances which suggest a more complex picture, including a more complex account of very elderly people.

Rather than older people being inevitably constituted as uninterested in physical intimacy and as sexless, there is wide-ranging evidence that sexuality remains a continuing and important aspect of older people's lives (see DeLamater and Sill 2005). Indeed, the 2015 *English Longitudinal Study of Ageing* (*Elsa*) notes that a 'sizable minority of men and women are still sexually active into their 80s' and that 'more than 41% of 80- to 90-year-old women who lived with a partner' report some sexual activity in the past year (Pidd 2015). And, internet dating apparently facilitates such sexual intimacies. For example, according to The Australian Seniors Series (2018) dating survey 'three-quarters of dating seniors believe physical intimacy is an important part of dating'. Intriguingly, three in five consider that such intimacy 'improves with age'—a point to which we will return. Additionally, seven in 10 dating seniors in this survey state that they generally kiss on the first date (2018). Malta's earlier study (2007) suggests that older Australian participants (between the ages of

60–92) in internet dating were not only swiftly affectionate if a first date went well but were also relatively quick to become sexually intimate with those they met. According to Noone (2012), Australian singles over 51, however they meet, 'are the most likely age group to have sex on the first date'.

Not surprisingly in this context, older people appear in a range of studies to challenge any assumption of age being linked to declining interest in bodily intimacy/sexuality (Meah et al. 2011: 62; Gewirtz-Meydan et al. 2018). For example, Frohlick and Migliardi (2011: 79–80) tell of 63-year-old woman, Jen, who may not want to 'show her boobs' in her profile picture but whose profile nevertheless says 'seeking sexy senior'. Jen explains that she would be happy to date younger men. While she may be looking for a long-term relationship, Jen is nevertheless keen to avoid 'boring widowers' (see also McWilliams and Barrett 2012: 426). Other over-50-year-old heterosexual women in Frohlick and Migliardi's Canadian study (2011: 83) report that they have enjoyed the opportunities internet dating has provided to have casual sex. Moreover, if given the opportunity, older people themselves express the importance of sexuality in living well. For Yvette (66) sex is 'very important'. She remarks that embarking on internet 'romances', in some ways. . .has opened my eyes to the fact that some men still find me quite 'yummy'. I like having that view of myself *(Yvette, Australia*—cited in Malta and Farquharson 2014b*)*. Similarly, a woman participant in Sue Malta's research on older online dating says, 'I can hardly walk, but there is nothing like a romp in bed to make me feel alive' (cited in Cooper 2008).[7] Such signs of bodily intimacy/sexuality are clearly at odds with the disembodied asexuality typically attributed to older adults. It would seem that internet dating provides a space in which to explore subversion of age normativity but, because ageist prescriptions intersect with gendered heteronormativity, it also has the potential to generate heterodoxical perspectives and practices that challenge assumptions regarding gender, hetero-sex, hetero-coupledom and hetero-intimacy more generally.

There are a number of indications of potential heterodoxies, such as older heterosexual women embracing options that subvert gender norms, such as beat-style, casual stranger-sex and inter-generational sex (see Malta 2007; Frohlick and Migliardi 2011: 83). Yet, relatively little academic research is available about older women's experiences and the extent to which they might use internet dating sites to facilitate sexual encounters. However, Frohlick and Migliardi (2011: 83) suggest that older women could use internet dating to 'get laid' in casual sexual encounters, and may indeed do so to keep their sexual activity separate from friends and families (see also Stephens 1976).[8] Moreover, it would seem that some 50+ women are not at all reticent about seeking to 'get laid'. This assertiveness is registered in comments from older men who are more than a little startled by their female peers' level of interest. Giles (68), for example, when he first began dating was flabbergasted by what he saw as their provocative sexual behaviour: 'I certainly didn't expect women to come on so strong' *(Giles, Australia*—cited in Arndt 2015b*)*. By contrast, others like Martin (69) express surprise but also delight.

> I guess one of the things . . . that I've been absolutely staggered [about] and you know, God, I'm no oil painting, but I've been staggered . . . how the middle-aged and mature woman is a very sexual individual who wants to go to bed and be stroked . . . and this, this surprised me.
>
> *(Martin, Australia*—cited in Malta and Farquharson 2014b*)*

While there is evidently considerable diversity in older women's sexual interest (Fileborn et al. 2015), Alison (66) provides one explanation for older women's enthusiasm in the light of her experience of a long period of abstinence after becoming a widow:

> It is all a big adventure. As long as I am up front with myself and my partners, I reckon I can do what feels good and have a ball. No more fears of pregnancy, no more of those crazy messages like 'He'll think you're a slut' or 'He'll think you are too easy'. What a load of rubbish all that was.
>
> *(Alison, Australia*—cited in Arndt 2015b*)*

Furthermore, there are signs that some older women are sexually interested in younger men (Alarie 2020; McKenzie 2015) and in pairing up with younger men online. Such connections upset the usual gender hierarchies in which older men select younger sexual partners who are traditionally deemed more attractive than older counterparts. The emergence of self-designated 'Cougar' dating sites suggests that older women are indeed taking up the opportunity to challenge gender norms by dating younger men.[9] Cougars are conceived as assertive, independent older women who actively pursue younger men and are sexually empowered. 'Older' when associated with the term 'Cougar' usually refers in popular and scholarly discourses to women who are 35 to 55 (Lowen 2020). Hence the category of Cougar barely rates as a substantive unconventional challenge to social norms which restrict older women. Most 'older' women, as we and many other scholars have defined them, are simply not included in the terminology, which largely appears to refer to middle-aged women. Yet, despite such limitations, the much publicised accounts of Cougars in popular and celebrity culture (Shpancer 2012; Weitz 2010) give credence to the existence of Cougar online dating sites/apps. And, in concert, these instances provide a language which at least opens up the possibility of older women's sexual agency—providing support for older women's (and men's) own increasing confidence and comfort in sexual expression (Rowntree 2014)—and may indeed in practice enable 50+ women to take up heterodoxical options in internet dating and elsewhere. For example, UK website *Toyboy Warehouse* requires its members to make match choices with an age gap, such that the average age difference between matches was reported as 12 years in

2016. Twice divorced 53-year-old Kate initially connected with younger men by accident rather than deliberate design but then embraced this opportunity.

> With older blokes showing no interest in me, I figured my time was up. So imagine my surprise to find younger men seemed to like me . . . We all have a fixed idea about what kind of person we'd like to meet, but beyond that, I treated the apps as a social experiment in the wide variety of people dating in the UK. If the chap texted in an articulate manner and wasn't a jerk, I chatted.
>
> According to dating site eharmony's research, the next decade's daters are going to be mainly in the 55–64 age range . . . And that's great if you're looking for someone in that range. I'm not.
>
> *(Kate, UK*—cited in Krizanovich 2018*)*

Other commentators have drawn attention to the difficulties faced by older women should they undertake relationships with younger men—difficulties Alarie views as arising because Cougar relationships 'violate long-standing social norms' (Bali 2021, Alarie 2020, 2019, 2017). They also paradoxically replicate some normative narratives associated with gender and sexuality. Nevertheless, we suggest that the apparent take-up of casual and intergenerational sex by some older women may be deemed a rather more significant departure from normative narratives than similar activities undertaken by younger women. Ageism's impact in relation to older women's intimate/sexual connections means that these indications of their sexual agency breach deeply held conventions, as Alarie's use of the language of violation signifies.

Additional heterodoxical directions in older internet dating include older women negotiating some control over the initiation and pace of dating and indications that older men may respond to the new experience of internet dating by shifting towards an expanded range of emotional and sexual communication (Rowntree 2014).[10] Older men in many studies report their reasons for dating as associated with their emotional needs. This orientation may be exacerbated by generally greater rewards and health benefits accruing to men than women from heterosexual coupledom (McIntosh et al. 2011: 70, 79). However, it is not simply a matter of these men expecting women to provide them with emotional and other services for their benefit. As Linn Sandberg's remarkable qualitative work on older Swedish men (aged 67–87) outlines, many of her participants assert that in later life their understanding of intimacy moved from an instrumental focus on their own pleasure in coitus towards a more mutual and communicative conception of relationships that is suggestive of movement away from traditional 'hegemonic' forms of masculinity associated with gender inequality (Sandberg 2013: 261–262; see also Levaro 2011: 115–121).[11] These additional heterodoxical directions are similar to those undertaken by younger people and rather than challenging social norms are perhaps more in the way of gentle modifications. All the same, these more unassuming modifications are seemingly more widespread, even common, and thus should not be underestimated. Social change may after all be achieved by many small-scale refashionings, what we have called 'micro-politics', as well as by fewer notable rebellions.

Heterodoxical activities undertaken by *both* older women and men mirror in many ways those of other age cohorts, such as resistance to marketisation, enhanced emotional reflexivity, rebuffing inequitable homogamy, and the growth of mundane polyamory. Nevertheless, innovations undertaken by older daters, though generally not intentional or ongoing, arguably offer more of a challenge to these existing norms than corresponding activities undertaken by young people because many such innovations also counter pejorative notions that older people are unattractive, not desirable, disconnected from bodily intimacy in relationships or asexual, and generally uninteresting and 'past it' (Minichiello et al. 2000: 259).

While most heterodoxical directions in older online dating may be deemed to involve modest departures from gendered heteronormativity, activities which we identified in Chapter 5 as troubling 'chemistry' appear to give rise to some further opportunities for subversive, even dissident, nonconformity amongst older daters. Those activities were described in the previous chapter as contesting gendered heteronormativity by delaying, bypassing or even resisting hetero-sex and hetero-coupledom. They involved diversion from online dating's central focus on the goal of romantic/sexual hetero-coupledom either into personal development or into various versions of platonic relationships. Such heterodox directions are not it seems limited to young daters. Older people also employ internet dating, along with other forms of online social networking, as a means to personal development (Erjavec and Fišer 2016: 366; Hogeboom et al. 2010; Jung et al. 2017; Parker and Carden 2018; McWilliams and Barrett 2012; Vandeweerd et al. 2016). In this context, Patsy (65) notes, after using online dating sites for eight years: 'I haven't minded [not meeting anyone] . . . I'm adding interest and variety to my life by having contact with people I wouldn't meet any other way. Anything else is a bonus *(Patsy, Australia—* cited in Alexander 2018*)*.

There is also ample evidence that many older people develop platonic relationships through internet dating. In Chapter 5, we paid attention to the development of emotionally intense alt-platonic/heterodox-platonic or queerplatonic (AP/HP, QP) relationships. These forms of platonic intimacies may also arise between older internet daters, but here we focus on the development of friendships and, in particular, cross-sex friendships. The significance of online dating as a means to friendship for older adults need not be assessed in terms of the presumed primacy and superiority of hetero-sex and hetero-coupledom, or even as somehow lesser compared with the more intense intimacy associated with AP/HP or QP relationships. Seeking out or finding cross-sex friendship is after all an unconventional option which remains largely socially unacceptable and usually treated with suspicion across the globe. Lapidos (2010) argues in this setting that cross-sex friendships are both profane and underexamined for a range of reasons, including disapproval of them as disguised or repressed sexual relations and hence as somehow deceitful and a potential threat to heterosexual partnerships. Relatedly, they are deemed unimaginable within societies marked by gender inequality on the basis that friendship requires equality, and are also presumed to be in a state of constant sexual tension which inevitably undermines friendship. Finally, cross-sex friendships are considered impossible on the grounds of gender difference such that hopelessly different types of people (men and women) are thought incapable of being friends (see also Chatterjee 2001; Deresiewicz 2012).

Lapidos (2010) asserts that male–female friendship is discomforting and typically taboo because it threatens the very basis of gender distinction and identities. We would add that this confrontation with gender inevitably also disturbs heteronormative heterosexuality. Cross-sex friendships as a common outcome or as an unconventional goal of online dating for over 50-year-olds thus constitutes at least subversion of gendered heteronorms or even dissidence in its more intentional departure from those norms. Although older women face a significantly more restricted dating pool than their male peers (linked to continuing male preferences for women younger than themselves and men's shorter life expectancies), they are frequently reluctant to re-establish a traditional heterosexual relationship. Older women are more inclined than men of their age cohort to prefer to remain single or to consider platonic relationships (McIntosh et al. 2011: 77–79). McWilliams and Barrett (2012: 415, 420, 423–424), among others, point out that for older women, online dating can provide companionship without loss of independence, hence offering novel heterodoxical models for connection and intimacy (see also EliteSingles Editor no date; Shaw 2020; Vandeweerd et al. 2016; Viveros and Schramm 2018). While older men in the same study by McWilliams and Barrett (2014: 424–425) are seemingly more interested in returning quickly to the social safety and personal support of exclusive romantic hetero-coupledom, they too frequently embark upon relationships which are not inevitably conceived in the same way as their previous marriage or marriage-like couple arrangements before they began internet dating. For instance, research on older Australians (60+) reports that, while romantic sexually intimate relationships are viewed as important, few lead to cohabitation or marriage (Malta and Farquharson 2014a; Erjavec and Fišer 2016: 362).

Cross-sex friendships are perhaps all the more unsettlingly heterodoxical because these relationships place in question the very essence of the gendered heteronorm and of internet dating itself, yet they are a common practice and feature regularly in older daters' accounts of the pleasures of online connections (Mansour 2017; Stephure et al. 2009). For instance, although Nigel (71) has clearly not resiled from pursuing 'romantic relationships', after using dating sites for several years he is very clear about his enjoyment in the cross-sex friendships he has gained.

> Obviously one has to be persistent and incredibly optimistic, but from my point of view I have had seven happy years, made some wonderful and permanent friends and lost nothing along the way. Where else can I get anything like that except on the internet?
>
> *(Nigel, Australia*—cited in Malta and Farquharson 2014b*)*

While the guiding premise of most dating sites is that the ultimate goal is love and marriage or marriage-like relationships, this ignores what matters to many 'seniors' who are looking for companionship and possibly even several companions to meet their varying social needs (Dowling 2014). While it is clear that most dating sites/apps 'are geared towards romance', this is not an adequate reflection of older adults' use of them since, for instance, The Australian Seniors and CoreData Survey 2016 shows that 'more than 40% of those surveyed have made new friends via dating apps and websites, and more than 20% would consider these people to be close friends' (Neeland 2019; The Australian Seniors Series 2016).

Yet, cross-sex friendships are not well acknowledged within the online dating services industry or even in academic research. Indeed, cross-sex friendships offer examples of much underestimated forms of dissidence. As Tarrant (2010), Calasanti (2009), Sandberg (2009), and Hearn and Sandberg (2009), among others, have noted in relation to older men, older people may provide unexpected examples of social innovation. We now turn to some further instances of practices which defy the conception of older people as narrow-minded, inflexible and hence not contributing to social change, let alone as providing leadership in social transformative practices.

The health-centric/medicalised gaze: Reimagining the deficit model

Ageism is associated with constituting later life as deficit and desexualised. We have focussed our analysis of heterosexual internet dating through the lens of *bodily* intimacy/sexuality to bring to the fore older people's full humanity by acknowledging their embodied sociality, their embodied connections with others. Our concern is to consider the potential in these embodied connections for contributions to social change. However, the deficit model of age is not only sexless but also has a very specific slant: it is characterised by the dominance of a biomedical perspective and an increasingly interventionist 'health' industry linked to the growth of pharmaceutical and surgical responses to ageing. This dominance means that not only are older people constituted in terms of being a health issue but ageing itself is now a pathology. Rather than ageing being a normal part of the normal human life cycle, it has been re-badged as a blight, fortunately capable of being at least slowed, deferred, even suspended with appropriate rational self-care in conjunction with medical assistance (Marshall 2010). This means that older people's engagement in internet dating and developing intimate connections with others are seen through the constricting gaze of deficiency such that so-called 'healthy' or 'positive ageing' (Hepworth 1995) becomes the other side of the same deficit-model coin. Whether ageing is negatively equated with decline or healthy 'sexy seniors' are promoted, these seeming opposed perspectives both render ageing as something to be disdained and kept at bay (Sandberg 2013: 264). This construction of older age disempowers those associated with it by offering no other path than evasion or rejection of it and thus limits the possibilities for heterodoxical alternatives which embrace and even celebrate ageing. We aim here to show how dating in late life brings into view other ways of understanding of older bodies, sexuality and gender that suggest contributions to social innovation.

King and Calasanti have observed that many in the 'West',

> equate old age with disease and decline, liken health to goodness, assume that persons can control their bodies (including brains) and health with diet or exercise regimes . . . They can medicalize signs of age and reject those people who appear unhealthy and old for not having altered their lifestyles to maintain visages of health and youth. To forestall exclusion and status, people often resist identifying themselves as old.
>
> (King and Calasanti 2013: 703)

Not surprisingly, in a context where many define older age as 'a social and medical disease' (King and Calasanti 2013: 704) many 50+ internet daters strive to attain the ideal of youth (Calasanti 2004), feel an obligation to appear looking young (Alterovitz and Mendelsohn 2009; Field 2018; Gewirtz-Meydan and Ayalon 2018; King and Calasanti 2013: 703–704; Marshall and Katz 2002) and approach online dating as a means to 'revitalisation' associated with youthful vigour including bodily/sexual vigour (Erjavec and Fišer 2016). They associate ageing with loss of one's gendered selfhood—with emasculation and de-feminisation, or even with no longer being a man or woman (Marshall 2008) usually perceived as manifested through loss of sexual agency (Ševčíková and Sedláková 2020). In short, the starkly ageist 'disease and decline' approach, which views older people in terms of gerontology and retains an emphasis upon biomedical understandings that underestimate social contexts and systemic power relations, continues even when the discussion turns to 'ageing well', or promoting 'positive', 'successful' or 'active' ageing (see for an early Australian example, Gong and Kendig 2016: 24; Minichiello and Coulson 2005). The quandary of such apparently contradictory accounts is that 'in fact both work to valorise youth at the expense and material realities of older people' (Fileborn et al. 2017) and both work against exploring how older people might actually engage in internet dating and in the process how they might contribute to social change. The biomedical approach to ageing shapes accounts of older people's quest to develop intimate connection with others precisely by valorising youthfulness and pathologising later life. This largely works to disempower older people, casting them as of little or no social significance, which has serious implications for examinations of 50+ online dating.

By focusing on the embodied experiences of older internet daters we can see that gendered heteronormative assumptions about them are by no means simply reiterated (Arndt 2015a; Camacho and Reyes-Ortiz 2005; Hickok 2017; Malta et al. 2020; Moreira et al. 2008). While many if not most academic studies and popular reports about older women daters remind them of the social/sexual difficulties and related health risks they face as they begin to embark on meeting men in later life, it would seem that there are other available perspectives—particularly from 50+ women themselves. Older women may face new experiences associated with dating with some trepidation, but post-menopause is not the end of novel possibilities and socially/sexually innovative responses. For example, Megan notes that after a lumpy start, a variety of unexpected opportunities came to light.

> How has my experience in dating changed since I started going through menopause? Well it's changed completely . . . The beginning of my menopause was the hardest bit. It was like I was possessed by another person . . . The second thing that changed was my libido. Anyone even touching me would make me feel ill . . . Also there was weight gain, so not feeling very sexy anymore, foggy brain, and this constant desire of living in my comfy pyjamas (sic).
>
> After starting to understand more about the menopause and my menopause symptom (sic) . . . I actually started getting my hormonal balance back a little bit (and also as menopause progresses it gets a bit easier) and I started to feel excited again about life again. I started to feel normal, like me again.
>
> And so I am now 53 . . . but for sure I feel more free, as I know exactly what I want and need. Regarding my sex life, it is different as well. It is harder to have an orgasm, and you don't get as 'wet' as before down there. What I realised is that you have to be true to your own body and know what makes you aroused . . . That's why after starting to use my lubricants everything changed . . . And for some additional fun, try putting some on your partner as well. Lubricants are amazing also if you want to play with yourself.
>
> Generally speaking . . . [i]f I was in a relationship, it will be for all of the right reasons, good chemistry with someone on my wavelength . . .

> But if for now I am single that's fine too, I'm in a good relationship with me, I am enough.
>
> *(Megan, UK, online blog, 2019)*

Megan's account certainly does not evade changes in her embodied self but also indicates these changes might be capable of negotiation and do not prevent her from registering substantive positive aspects of ageing including a growing sense of freedom from social restrictions and of her own autonomy. One consequence of these positive elements is that sexual relationships with men are not the only or the most important options in her life. As pointed out earlier in this chapter and in the previous one, older women show flexibility with regard to dating which trouble traditional conceptions of hetero-femininity and the supreme status of hetero-sex and hetero-coupling. Developments like casual sex, sex with younger men, alt-platonic/heterodox-platonic or even queerplatonic (AP/HP, QP) relationships, and cross-sex friendships, as well as greater awareness of their own bodies and desires alone and in company, become imaginable.

The medicalised emphasis on low libidos and atrophied dry vaginas in much of the scholarly and popular culture concerning older women and dating (on or offline), offers pharmaceutical and surgical rejuvenation associated with hormone replacement, testosterone therapies and, increasingly, cosmetic surgery. Such a perspective involves comparing older women with their younger counterparts and on this basis reinforces uncertainty about the former's attractiveness and desirability in gender normative terms, along with doubts about their desire and capacity for heterosexual sex—equated with penis–vagina intercourse (Loe 2004: 322; Marshall 2008: 25; Sandberg 2013: 264). By this means, men's penises and erections are often, dare we say, re-erected as central.

Older women (along with older men) may be cautious, even possibly intimidated by beginning to date potential new partners and perhaps entering new intimate/sexual relationships, but they do not necessarily display particularly marked concerns about their bodies. Indeed, 'women aged between 65 and 74 tend to be the most confident about their bodies when compared with other age groups' *(Anonymous, UK, dating site advice blog, 2019)*. Moreover, Megan is not alone in finding that dating enabled her to recognise and develop forms of confidence precisely associated with being an older woman. In a similar vein, Caterina, a '50-something divorcée', discovered after embarking on internet dating that 'ageing is a privilege, not something to dread'.

> I knew my 28-year marriage was over . . . My passion for an emotional, romantic life was spent. Desire would play no part, I decided. I would be celibate . . . The last time I'd been on a date with a man other than my ex-husband was 1984 . . . I finally took the plunge [and] chose Bumble because I liked the idea that women make the first move . . .

> Post-menopausal women are actually quite a desirable lot: more confident than our younger sisters, clear about what we want and even attractive to younger men who like the fact that we're more comfortable in our own skin, even if that skin is a little crepey. Most of my dating friends have been out with younger men and in some cases substantially younger men . . .
>
> My friend L maintains the only purpose of dating apps is to stop hoping you'll meet someone to complete your life every time you go to dinner or a party. For me it has affirmed that my life is already a good life. I'm looking for the cherry on the top of a cake I chose and baked myself. I don't need it but it would be fun.
>
> Dating has given me a chance to laugh, be carefree and connect with something I'd lost in myself . . . Maybe dating in your 50s isn't about love, companionship, orgasms or even passionate kisses. Maybe it's about feeling that your life is still opening up, not shutting down, and that you can choose which direction you want to take it in. Online dating could give you confidence both professionally and personally, and learning how to flirt again isn't a bad thing either.
>
> *(Caterina, UK, online blog, 2018)*

Older women's embodied practices evidence heterodoxical innovations from minor divergence to dissidence. Menopause and beyond do not apparently inevitably signify either dispiriting decline nor the emergence of the age-defying 'sexy senior' but rather a more complex picture which constitutes ageing as precisely enabling new social directions that are perhaps less available or even not available to the young. To entertain the notion that older women, rather than being left behind in the back blocks of history, might be the vanguard of societal transformation in relation to gender and sexuality is indeed a challenge to commonly accepted ideas of age, gender, heterosexuality and social change.

When we turn to older men the limitations of a biomedical agenda again become obvious. As has been intimated in discussion of older women, the reduction of older men's social and sexual health, intimacy and pleasure into sexual 'functioning', defined by the desire and ability to engage in penis–vagina intercourse and relatedly to have erections, effaces registering other ways of seeing masculinity and age (Fileborn et al. 2017: 2098; Hughes 2011; Marshall 2008: 22; Sandberg 2013: 264). Rather than erectile 'dysfunction' dominating understandings of older men's negotiation of bodily/sexual intimacies, it would seem that the experiences of these men may challenge gendered heteronorms in relation to an increased regard for emotionality and communication with female partners, a considerably expanded range of understandings of pleasure and of sexual practices, and an accompanying sense not inevitably of loss but of the advantages of ageing (DeLamater and Koepsel 2015; Menard et al. 2015; Sandberg 2015). Sandberg (2013: 267) describes this overall shift as moving 'from the "fiery urges" of youth to the freedom of intimacy'. This is not a deficit model or in any sense one which replicates the equation of proper manliness (associated with hegemonic masculinity) with strength, competition between men and subordination of women (Beasley 2015, 2008b; Calasanti 2004; Lodge and Umberson 2013). Rather, Sandberg's older male participants experienced intimate relationships as 'taking new shapes' away from a goal-oriented hetero-sex focussed upon men's performance and orgasm towards more open-ended and interactive expressions of intimacy which they often described as enabling them to become 'better, more considerate lovers' and as having positive consequences for their relationships with female partners (Sandberg 2013: 271). Such accounts do not appear to be entirely self-serving since Lodge and Umberson (2012) report that mid-to-later life couples note a decrease in frequency but an increase in the quality of sexual encounters. Similarly, participants in Menard et al.'s 2015 qualitative study assert that their sexual experiences had improved in quality over time and involved developing skills associated with knowledge of one's partner. These earlier studies add weight to the findings of The Australian Seniors Series dating survey (2018). As noted earlier in the chapter, three in five respondents to this survey—including women as well as men—assert that physical intimacy gets better with age.

Older men frequently feel disconcerted and even express dismay when faced with the possibility that they can no longer rely upon an erect penis that they have previously assumed to be sufficient unto itself as the synecdoche for the sexual desire and pleasure of both men and women. Yet, loss of certainties around erections, and associated doubts regarding phallocentric social norms, might not simply be deemed a misfortune and a medical 'problem'. Castleman (2012) states, following landmark studies by the University of Chicago, that more than half of men over 50 do not develop erection 'dysfunction'—that is, the inability to have an erection even after extended masturbation (see also Laumann et al. 2008). Indeed, ageing does not inevitably produce dysfunction, so much as 'erectile dissatisfaction' when compared to youthful vigour. This dissatisfaction amounts to the acceptance of ageist presumptions in which youth is taken as the axiomatically superior and standard against which all other adult age cohorts should be judged and found wanting. It is fuelled by both a conception of manliness as located in the penis and its 'performance' such that being forever hard and hence ready for intercourse is a requirement, and a 'sexual health' industry eager to reinforce this conception (Marshall 2008: 22). In this setting, older male daters sometimes report that they found it difficult to acknowledge difficulties with erections, but nevertheless many also note with gratitude and surprise that older women were not always as disturbed by this as they had imagined would be the case (Price 2015).

However, whether older men are able to have erections easily or at all, many discover that erections are not necessary for sex, sexual enjoyment or orgasm. As one of the participants (in his 60s) in Malta et al.'s Australian research examining older adults' conversations about sex in primary healthcare settings makes clear, what is deemed to be 'sex' is subject to change. Over the life course, he suggests, it is not necessarily predicated on intercourse: 'you can do other things and pleasure each other in other ways [which] can actually create the intimacy that can be lost if you're not having . . . penile–vaginal [sex]' *(Marco, Australia*—cited in Malta et al. 2020*)*.

Although this shift may still involve certain features of hegemonic masculinity in attending to sexual mastery and skill and hence potentially reiterate gendered heteronorms, Fileborn et al.'s 2017 research shows a consistent and strong concern by older Australian men to stress mutual enjoyment and equality as making sex pleasurable (2017: 2106; see also Hughes 2011). For Berndt (71 years old), in later life sexual pleasure 'turns into more like stroking or a comfort thing, asserting that your partner is still a valuable person'.

Sexual interactions which do not reply upon having or maintaining a hard penis may be longer and slower, involving a greater need to communicate one's desires and requirements, but they may also produce enhanced intimacy and sexual enjoyment for heterosexual couples. More specifically, Lodge and Umberson (2012) suggest on the basis of their American research that as men move from mid-life (50–69) to later life (70–86) they are increasingly likely to emphasise the importance of emotional intimacy and to experience greater congruence with their partner's experiences of sex. In this light, for many older people, 'erection dissatisfaction can be a gift' (Castleman 2012). Rather than a soft penis necessarily representing loss and decline, a renewed focus on touch, fondling, hugging, kissing, mutual or parallel masturbation, oral sex, sexual 'aids' and sex games offer a novel prioritising of intimate sexual practices normatively cast as mere 'foreplay' and/or as bonus benefits rather surplus to the requirements of the supposed main event of penis–vagina intercourse. This alternative view suggests quietly subversive, even dissident practices enabling an unshackling from the constraints of normative heterosexuality. Moreover, it is important to keep in mind what Gayle Rubin calls the 'fallacy of the misplaced scale'—that is, when sexuality scholars assume that sexual activities have a greater importance than other human activities and accordingly give sexuality an unwarranted 'special status' (Rubin 1984: 278; Jackson and Scott 2004). There is a danger that in countering ageist assumptions depicting older people as asexual beings, those older adults who do not problematise loss or lack of sexual interest or practice may be othered or ignored (Sandberg 2015: 4). Older men's experiences—whether they involve sexual desire and sexual activities or not—may provide creative heterodoxical models for connection and intimacy as well as for hetero-sex and hetero-coupledom.

As has been pointed out in relation to older women, rather than older men being 'past it' and hence of little interest in examinations of sociality and social change, an alternative perspective comes into view. While young men may be heavily policed to aspire to and perform forms of hegemonic masculinity that replicate men's authority over women through reiterated practices associated with the centrality of the penis and its companion, the 'coital imperative'

(Jackson 1984; Willis et al. 2018; see also Chapter 3), ageing men it appears may undertake new social/sexual directions. Men over 50, along with their female counterparts, indeed can offer us intimations of more equitable and tender—though none the less intensely felt—relations between men and women. Older men too may defy expectations and unexpectedly be the avant-garde, at the forefront of social change, when it comes to gender and sexuality.

Creative trailblazers are not always or inevitably youthful but instead it seems may be mature or even elderly. While some scholarly studies acknowledge that later life is not all doom and gloom, promote conceptions of older people's full humanity and even occasionally observe intriguing signs of life in them—particularly in the baby boomer generation, to our knowledge there is no research which as yet recognises that older people might in practice constitute the cutting edge of social transformation and represent unacknowledged and under-theorised modes of liberation from normative gender and sexuality. In this context it is useful to recall that when Michelangelo turned 70 he had a further 19 years to live and in those years produced some of his most powerful works. While 'David' was a work of his youth, Saint Peter's Basilica and the 'Bandini' Pietà were productions of his old age. As Ingrid Rowland (2020: 8, 10, 12) points out, 'we have a tendency to ascribe creativity to youth itself', but mature virtuosos like Michelangelo remind us that the capacity to display creativity 'depends on age to reveal its full complexity'.

Ageism, inflexibility and social quietism? Concluding alternative perspectives

As we noted at the beginning of this chapter, being old is almost invariably linked with a range of unappetising characteristics (Minichiello et al. 2000: 259) which not only express distaste but perhaps even more importantly suggest that older people are without interest because they are seen as looking backwards rather than forwards *(Laurel, UK, online newspaper story, 2016)*. They are thereby linked to closed-mindedness and social quietism. However, as this chapter shows, older adults should not be ignored in social analysis. Their engagement in online dating illustrates a variety of examples of social innovation. There is much to indicate that far from being immured in social quietism, they are involved in and contributors to change.

A blog attached to an Australian dating website for older people suggests that age actively adds to the capacity of older internet daters to undertake innovation. In response to the question, 'Can older dating help you face change?', one blogger is quite clear.

> Older dating with other senior singles . . . can provide you with a support network of friends and a special senior single partner to help you through these changes, support can be really essential when your life is in a state of flux and within the senior single community of [dating site] you are bound

> to find others like you who have 'been there and done that'. The very nature of older dating means that the majority of senior single members have faced major change!
>
> *(Trevor, Australia, dating site blog, 2014)*

This contribution to a dating blog is a way of promoting the site and encouraging people to join. Yet it also claims that adults in later life do not necessarily face 'fewer major life changes' as they grow older and that internet dating may be both a positive factor in adapting to those changes and demonstrate how older adults have proved able to face and embrace change.

Internet dating demonstrates much that remains normative. Nonetheless, our focus upon older people and bodily intimacy/sexuality reveals many heterodox innovations which are politically significant when considering social change with regard to gendered heteronormativity. Fifty plus daters are not inevitably frail, senile and unattractive, let alone boring, as their age is supposed inevitably to imply, but rather an unrecognised resource for reimagining social and sexual life.

Notes

1 This chapter arose in part out of earlier works and particularly draws upon the first mentioned of these (Beasley and Holmes 2016; Beasley 2015, 2011).

2 While there has been a rise in designated dating services for older people and many dating apps and sites have sought to differentiate themselves in ways that are deemed to be more appealing to an older cohort, it is also the case that *Tinder*, for instance, is not especially older-friendly and has higher charges for users over 30 (Dolan 2020).

3 Paisley concludes (research findings draw upon 11,425 daters across the globe with the exception of the Middle East) that 75 per cent of online daters are aged below 30 (2018), with much higher overall usage in Latin America and parts of the Asia-Pacific. Dolan's 2020 assessment, which is inclusive of the Middle East, reiterates this general survey noting the markedly greater percentage of internet daters located in the Asia-Pacific region (see also Chapter 7).

4 Here are just a few recent examples of the different age associations given to 'older' in relation to internet dating (McWilliams and Barrett 2012–53–74; Mershon 2019—over 60; Wada et al. 2016—over 60; Wion and Loeb 2015—over 50; Hill 2020—in the article over 50 or, in the title, over 60; The *Telegraph—Dating* 2020—over 65).

5 We conceive of intimacy and sexuality as located within a set of associated concepts but recognise that not all intimacy is sexual and not all sex involves other forms of intimacy (see Chapters I and 5 for more detail characterising these concepts).

6 The sexually transmitted disease (STD) health risks so often the subject of concern with regard to older people's take-up of internet dating may be reasonably deemed 'overblown', in that few women over 50 contract chlamydia, there are no risks regarding pregnancy, herpes risks recede and HIV, syphilis and gonorrhea risks are small. Once again, it seems that anxieties concerning health are strongly foregrounded whenever older adults are the topic of discussion (Arndt 2015b). Ageing is almost inevitably associated with health fears.

7 The quotation is reported by Cooper in advance of Malta's presentation to the Australian Sociological Association conference in December 2018. Access to that presentation is not yet available.
8 For example, as Frohlick and Migliardi (2011) and Stephens (1976) amongst others note, it is not uncommon for adult children to constrain their parents' engagement with dating seemingly because of its discomforting intimations of parents having a continuing interest in sex.
9 There are a range of definitions of the term, 'Cougar', but most refer to middle-aged women who date or wish to date significantly younger men—for example, younger men who are ten years their junior (see Lawton and Callister 2010: 6–7).
10 Such shifts by men towards greater emotional reflexivity are evident in other relationships that depart from the norm (Holmes 2015).
11 Within the field of Critical Studies in Men and Masculinities, Raewyn Connell's terminology of 'hegemonic masculinity' is virtually omnipresent and is a means to recognising that not all masculinities are equal. Rather, hegemonic masculinity holds an authoritative positioning over other masculinities, offering an honoured and mobilising model of manliness to which all men are supposed to aspire. It is however crucially defined by its function in legitimating gendered power relations, specifically men's power over women (Beasley 2008b: 88–89; see also Beasley 2012).

Chapter 7

In an international frame

In this chapter, we explore novel modes of developing intimate relationships arising within the growing use of internet dating sites worldwide. This technology of intimacy offers a means to meet social norms regarding gender and sexuality with regard, for example, to fulfilling societal requirements such as limiting cross-sex contact and premarital sexuality and encouraging monogamous marriage, yet simultaneously it may enable some avenues to redirect these norms away from parental, familial or other social restrictions. Our aim in the book as a whole and in this chapter is to highlight such heterodoxical possibilities.

Nevertheless, the seemingly open-ended scope and flexibility of internet dating is by no means unlimited. While Tinder is specifically a proximity-based model for meeting other people, dating sites/apps generally involve 'intimate geographies' (Pratt and Rosner 2006: 21; Walton-Roberts 2010).[1] The language of 'intimate geographies' enables intimacy to be viewed not simply as an individual or even societal matter but as also produced by place and space. This approach to intimacy shows the significance of an international frame in understanding internet dating and contributes to our rationale for this chapter. We refer to intimate geographies to draw attention to the ways in which dating sites/apps typically give internet daters more efficient (in terms of supply of potential contacts) access to *locality-based* dating pools. In other words, intimate geographies for the most part continue to shape daters' likely access to potential contacts in terms of acceptable physical proximity, language usage, religious/ethnic and other affiliations.

We would add that scholarly research on internet dating is typically also geographically localised. This limitation is particularly problematic in the context of the extraordinary dominance of American sources in the field. A review of academic publications from 2005 to 2019 substantiates this virtual monopoly and shows the lack of studies from many countries around the world (Hong and Lia 2019). As noted in Chapter 1 and elsewhere in the book, we have focussed our attention upon the UK and Australia in an effort to provide some alternative contextual material which can be considered against the extensive body of literature from North America, and in particular the

United States. By this means we aim to foreground that internet dating is by no means all of a piece. This point is especially crucial given that the central question which drives our endeavours in this book is to examine whether internet dating can be a site for social change. Any response to such a question must take account of the extent that this relatively new technology of intimacy is unevenly researched, even though responses to it are likely to be culturally varied. Existing studies of internet dating are restricted in *scope*—in terms of their geographical and cultural narrowness, as well as largely giving little attention to social innovation. This means that present scholarship is also limited in *perspective*.

Developing an international frame

In order to expand the viewpoint of research on internet dating, we undertook a modest pilot study which once again involved our central concern with UK and Australian heterosexual internet daters but involved a collaboration with two other colleagues who focussed their investigations on Denmark and Uganda.[2] Our research thus examined dating sites and profiles in four countries.

This was a meta-level analysis, not including any identifying details, of public profiles on dating sites. It examined presentations of self with regard to appearance, attitudes, practices and relational connections. Furthermore, the Denmark and Uganda studies involved posting a question to chat sections or Facebook groups related to some of the dating sites asking if users could identify any new or different experiences associated with internet dating compared with traditional offline experiences. The collaborative team of four authors published an account of the pilot study in 2018 (Beasley at al. 2018).

Following this collaboration, in this book we two authors embarked on some additional research. We sought out academic, journalistic and popular accounts, as well as undertaking a more limited, schematic examination of sites and profiles for illustrative instances in South America, Asia and the Middle East. The schematic study was carried out for this book between August and September 2020. The profile materials from the four countries that were the sites of the collaborative pilot study (Beasley et al. 2018), in combination with instances drawn from Argentina, China and Turkey in the more restricted schematic study in 2020, together provide more in the way of evocative suggestions as against the detailed analysis of UK and Australian internet daters we have fleshed out in the larger part of this book. However, the broadly suggestive accounts of dating sites and profiles provide another perspective that can usefully be read in conjunction with the more comprehensive analysis obtained from the UK and Australia. While we offer some international observations regarding heterodoxical innovations in gender and sexuality derived from this selection of seven countries, representing the major world regions, the discussion of possibilities for social change arising in different geographical/cultural contexts is more in the way of juxtaposed observations than an extensive comparative analysis of such possibilities across the globe. Further research on this topic is much needed.

Our international perspective is limited in certain ways, nonetheless the use of a feminist framing gives a distinctive critical flavour and depth to our examination of heterodoxical innovations in gender and sexuality which augments the viewpoint of our research. Such a framing enables concentration upon axes of power relations while also enabling consideration of subtle, often unassuming nuances in relation to innovative intimacies and their transformative potential (see Beasley et al. 2017). An important aspect of this feminist framing arises in its emphasis upon an alternative methodology involving attention to the micro-politics of social life, including consideration of the everyday, embodied, emotional, libidinal and personal aspects of private conduct. This feminist orientation has theoretical, empirical and strategic advantages when examining heterodoxical intimacies within a variety of cultural settings. As noted in the Introduction (Chapter 1), the foregrounding of the significance of power relations and the micro-politics of social life enables a conception of social change which cannot be assumed to be only top-down. A central concern with micro-politics makes looking at practices of intimacy like dating a means to understand shifts in social life.

Pilot study materials and methods: UK, Uganda, Denmark and Australia

Our collaborative pilot study offers a comparative examination of certain dimensions of internet dating across several continents. The aim here is to offer snapshots of indicative directions regarding the significance of cultural context in the development of social innovations regarding gender and sexuality, before turning towards more broad-brush observations (in the schematic study) arising from additional cultural sites.

Our methodology with regard to the collaborative pilot study of UK, Denmark, Uganda and Australia requires some brief exploration insofar as it is a content analysis, thus differing from the detailed focus on the voices of internet daters in earlier chapters. Content analysis encompasses varied stances towards and means of analysing textual materials, including visual texts (Cavanagh 1997). These stances range from impressionistic to strictly calculable, from a qualitative stress upon the complexity of language, context and meaning, to a quantitative focus on manifest content and word frequencies (Rosengren 1981). The method employed may be 'prescriptive' and attentive to a closed set of parameters, or 'open' and intent upon bringing to light dominant and reiterated, as well as innovative themes (McKeone 1995). It may involve coding arising from the textual data ('conventional'), from pre-existing theoretical frameworks or research findings ('directed') or from key words (or themes) interpreted through the lens of context ('summative'; Hsieh and Shannon 2005). Images might also be analysed (for example, Goffman 1987). The preliminary status of this pilot study research, and the contextual variability of its multinational scope, necessarily situates our methodological approach towards the impressionistic and interpretative end of the content analysis continuum. Furthermore, our research and theoretical interests led to an 'open' emphasis on themes (both dominant and innovative) in keeping with a qualitative concern with meaning and intentionality, rather than making use of a strict, more closed concern with specific frequently repeated words. The overall shape of the study was framed by pre-existing theoretical research work around possibilities for heterodoxical innovation ('directed' approach), but we also made efforts to allow for a high degree of contextual sensitivity associated with cultural/national variation ('summative' approach).

The different contexts affected the material collected, especially the differing level and ease of internet access within the countries included in our study. For example, Okeleke stated in 2018 that 'significant digital . . . gaps exist' in Uganda: 'around 4 in 5 people, including more than half of the adult population, remain offline (2018: 8; Kanaabi 2020). To this may be added high levels of income inequality and marked gender inequality (Okeleke 2018: 9; ranking 159 out of 189 countries in the UN Development Program Gender Inequality Index GII 2019). Within this digital and social setting is a young, relatively economically privileged group of individuals who have turned to the internet for building intimate relationships. By contrast, Britain, Denmark and Australia have broadly similar and much higher levels of usage. This is but one difference that makes the comparison valuable.

We sought differences in the four sites that might be relevant to considering the impact of cultural context on possible social innovations. The contexts we chose are ones in which the collaborators were either working, and/or are from; two of the collaborators undertook research in countries in which they are employed while two embarked on research in their countries of origin. This meant that we specifically did not require that the several researchers in the study team should code in exactly the same way, as rigid requirements for inter-coder 'reliability' (see for example, Weber 1990: 12) were seen as problematically limiting responsiveness to context.

The collaborative pilot study involved parallel examination of profiles on particular sites in each country and was carried out in a preliminary way between September and December 2014 in the UK and, in concert by all the collaborators between 15 December 2015 and 15 January 2016. The four of us examined a sample of profiles in each country, that was purposefully attentive to gender, and in two studies, to age cohorts. In short, the collaborative team selected profiles as randomly as possible given the ordering sometimes imposed by the sites, until we arrived at equal numbers of profiles by men and women. In the UK and Australian research, we also undertook to arrive at a sample containing a range of adult age groups or paid specific attention to particular cohorts. The pilot project gave close attention to 80 profiles (40 men and 40 women overall across the four designated countries —although the UK research analysed 120 additional profiles of 25–45-year-old men). In the case of the Denmark and Uganda studies, collaborators also asked internet daters for direct responses to a query regarding the nature of their experiences. Sometimes more than one dating site and additional profiles in a particular country were included. Each collaborator selected 20 profiles to contribute a mix of 10 men and 10 women users from their chosen internet dating website or websites. The examination of profiles focussed upon meta-level patterns such as the most popular descriptors and thus is completely anonymised.

Pilot study: Contextual snapshots

In this section of the chapter we aim to give a sense of notable characteristics of online dating in the four countries chosen for the pilot study—which expands on remarks in the previous collaborative publication (Beasley et al. 2018), and to be placed alongside marked features in three other countries. The four country-based analyses from the pilot study take the form therefore of snapshots. We provide a brief background on each country outlining major feature/s relevant to and about internet usage and internet dating usage, followed by a condensed account of the differing collection methods we employed in the differing sites and a succinct account of an exemplary dating site (or sites) and associated profiles. Finally, we offer some assessment of consequential signs in the four settings with regard to both reiteration of and resistance to gendered heteronorms.

Plenty of Fish and Match.com, UK

The United Kingdom is a developed country, holding the position of the 21st most gender equal nation in the world at the end of 2019.[3] It is relatively small with a highly urbanised population (Worldpopulationreview 2020; Statista 2020c; Plecher 2020; Worldpopulationreview UK-cities 2020; Rae 2018). The union of England, Scotland, Wales and Northern Ireland within the UK has produced a stable democracy with robust legal protections for political and civil rights, as well as a well-established welfare state. Though Britain includes a mix of ethnic communities, 'it is overwhelmingly white' and only a small percentage of its population do not speak the official language of English (Platformglobig UK 2020). The UK has a 'very high level of income inequality compared to other developed countries and is even more unequally divided by differential wealth'.[4] Additionally, there are increasing concerns regarding the rise of racism and antagonism to immigrants, as well as regarding the use of technologies to enable surveillance of citizens (Freedom House—UK 2020c; The Equality Trust 2019).

In keeping with its developed economy, the extent of internet penetration in the UK is very high: the share of households with internet access was at 93 per cent in 2019 (Johnson 2020). Britain's substantial access to information and communication technology (ICT) is registered in its IDI score. The IDI is an index which, based upon a range of indicators, measures the digital divide across the globe comparing ICT performance within and across countries (ITU 2020). The UK IDI ranking (ITU-IDI 2017) is fifth in the world, demonstrating high levels of, for instance, access, use and skills.

The social acceptance of internet dating usage in Britain is evident in the continuing rise of users registered on sites/apps (Belton 2018). Fifteen million singles in the UK are reported as registered for online dating. They currently have over 1400 dating websites to choose from. *Match.com, UK* a comparatively large site is said (at 2020) to have 10 million subscribers (Clement 2020; Visual.ly 2020). *Plenty of Fish (PoF)* is one of the more affordable sites.

The analysis of UK dating profiles entailed 20 profiles (10 women, 10 men) on the site *PoF* and 120 men's profiles on *Match.com*. When loading *PoF*, about five profiles appear at a time. Creating your own profile is necessary for users wanting to contact people via the site, but basic profiles can be viewed by anyone. These public-level profiles mostly belonged to users in their 20s and 30s, with fewer over 40. We went through the profiles, selecting examples in the order they appeared, to achieve an even spread from four age groups (20s, 30s, 40s, 50s), until 20 profiles (10 men, 10 women) had been chosen. Eighteen users identified as white, two as black. Limited information about level of education and current employment was available. Clicking on a profile lead to an in-depth page with basic details: name, description, city, information such as gender, height and religion, ethnicity, intent (such as, 'wants to date but nothing serious'), education, personality (for example, 'daredevil') and profession. There is space for a profile photo, which is followed by a text-based section requiring short responses to statements (like, 'I am seeking a . . .'), as well as information regarding marital status, personal characteristics like ambition, pets, and so on. Last, there is room for a paragraph in which the person writes about themselves and what they seek. A further content analysis was conducted between September and December 2014 on *Match.com, UK*, a dating site which requires people to pay a fee if they wish to exchange messages. As in the case of *PoF*, it is possible to register for the site and view profiles without paying. However, in this case a profile named 'Researcher' was temporarily put up on *Match.com, UK* and men 25–45 years were searched. The profiles were selected and analysed as for *PoF*.

In considering the UK sample of profiles on both dating sites, one of the most notable aspects of these accounts of self is the tension between gendered heteronorms and discernible if unobtrusive signs of challenges to these norms. There is evidence of reproduction of normative ideals of 'emphasised femininity' (Connell 1987: 183–190), and—for the most part—of rather traditional presentations of masculinity. On *PoF*, women's main profile photos are mostly headshots, several of them pouting suggestively. All wear make-up and most have long hair. Women often describe themselves as 'divas', fashionistas or princesses, with one or two using terms like loyal or honest.

It appears that men's representations of selfhood on the two British dating sites were not obviously strongly normative in terms of 'hegemonic masculinity'. While the visual markers men use in presenting themselves online may in recent times increasingly put emphasis on male bodies deemed normatively attractive in hetero-masculine terms (Allain and Marshall 2017; Novak 2014; Schwartz et al. 2010; Siibak 2010; Stevens and Ostberg 2012; Wahling et al. 2018), this was not especially evident in this sample. Proper manhood is something still strongly associated with many contemporary Western forms of hegemonic masculinity in which being a (strong, silent) man is an obligatory aspirational performance and in itself unquestionably valued (see for example Beasley 2008b; Connell 1987; Connell and Messerschmidt 2005; Kimmel 2004; Perkins 2015: 9–14, 29; Ricciardelli et al. 2010: 65). The normative reiteration of proper manhood in the visual imagery chosen by men for their dating profiles as a means of attracting potential female partners does not, initially at least, appear highly gender stereotypic. Nonetheless, gendered representation is still deeply embedded in the profiles (cf. Goffman 1987).

Men's photographs on the UK sites were mostly headshots, but many were poorly taken 'selfies', showing a certain disregard for their appearance. On *Match.com* there are fairly evenly balanced numbers of photos taken indoors and out, with only some men doing active, sporty 'manly' things (Bryson 1987; Connell 1987; Drummond 2002, 2010; Fitzclarence and Hickey 2001), and a few men pictured with cars (Balkmar 2012; Duerringer 2015; Walker et al. 2000). On *PoF*, the men's main profile photos are more likely to be torso shots. Only one or two of these men were topless, with one wielding two surfboards, yet their bodies were not especially on show in relation to appearing sexy or buff. While conventional portrayals of men as active and sporty and/or muscular still appear on the site (with many men in this sample wearing sunglasses—perhaps another marker of outdoor activeness), the men's photos on their profiles remain decidedly gendered in the more old-fashioned, though perhaps somewhat less obvious, sense of a seeming unconcern with how they look.

By comparison, women's profiles show a decidedly greater concern with presentation in terms of clothing, hairstyling and cosmetic enhancement as well as in relation to organising photos that are at least somewhat professional looking. Women's profiles also show considerably more attention to conjuring up a persona through descriptive narrative. We see here a marked gender differentiation in these dating profiles involving an apparent reiteration of the traditional conception of manliness as, unlike femininity, disdainful of appearance and communication as not relevant to masculine desirability.

There are, however, some muted signs of resistance to gendered heteronormative ideals of relationships evident in the words used by daters to describe themselves. Despite the persistence of gendered heterosexual norms regarding activities, appearance and narrative expressiveness as indicators of femininity and manliness, these descriptors sit in tension with other directions in the profiles.[5] For example, a few women's profiles challenge normative understandings of women as dependent on men (Kittay 2013). They describe themselves as 'independent' and/or mention that they work and have their own home, although in this sample this is mostly a feature of profiles of women old enough to have gained some financial autonomy (later 30s and 40s). Men's profiles were more likely than those of women to contain references to being athletic or liking sports or the gym. Several men said they are not looking for commitment or anything serious and two included no descriptive information. These profiles seem to conform to the stereotypically expected trope of men as looking for casual sex online and not being interested in developing much exchange of personal information with women through their profiles. What is striking across both dating sites is how common the descriptor 'easy-going' was for the men. 'Adventurous' and 'funny' were the nearest contenders, although 'fun' was also commonly used in the *PoF* profiles. It is a tantalising question as to whether such terms are employed by male daters to imply that they are not pushy, aggressive or hyper-masculine and whether such descriptors indicate men's awareness that such characteristics are not likely to attract interested women.

Women in the sample undertake innovations in hetero-gendered femininity by using their profiles to signal their intolerance for forms of hyper masculinity. They also filter out or screen men who are seeking casual sexual encounters, and/or are overly familiar, sexually aggressive, inappropriate or sexist (see Chapter 2; see also McWilliams and Barrett 2012;). A few of the women's profiles, for instance, contain pleas for men to only contact them if seriously interested in a relationship and asking them not to send 'knob shots' (photos of their penises), and not to propose casual sex, an affair, or to call the women 'luv' or in other ways 'waste' their time. These assertions may in part be understood as women simply replicating their traditional offline feminine role as sexual gatekeepers and/or harnessing increasingly widely accepted social narratives of respectful mutuality associated with hetero-coupledom in Britain (Jones et al. 2019; Taylor and Scott 2018: 56). Nonetheless, there are undoubted indications here of a refusal of notions of women as subordinate to men or as without rightful social agency.

The UK may be viewed as a society with a reasonable degree of gender equality in world terms. Yet, profiles on internet dating sites appear to offer few explicit signs of social change in relation to the intimate domain of heterosexual gender relations. All the same, there are hints in these profiles of subtle changes in interactions between men and women such that men often aim to present themselves as affable and less than traditionally hegemonic while women sometimes state their resistance to traditional conceptions of manliness. These representations of the self may indicate fine-grained fissures in the supposed monolith of gendered heterosexuality, particularly when placed alongside evidence of numerous heterodoxical practices we have outlined in previous chapters. The fact that profiles are like preliminary advertisements aimed at attracting attention from the opposite sex might make them the most averse to change within online dating (see also Chapter 3). Other aspects of internet dating which involve interactive practices requiring engagement and connection with potential partners (such as the social netiquette attached to messaging in dating sites) seem more amenable to alterations in social conventions regarding gendered heteronormativity. We should not assume that all elements of sociality, in this case within online dating, are equally responsive to innovation, and innovations will in any case vary depending on context.

Mingle2, Uganda

In the Ugandan context, internet dating is impacted by the country's significant income, ethnic and gender disparities as well as by its primarily rural agriculturally-based population, and its youthful demography. Although Uganda has about the same land area as Britain, it has around half the population, and the great majority of residents live in rural areas, a high proportion of people are in younger age groups and life expectancy is low (Among and Munavu 2019; Macrotrends 2020; PAI 2010; PopulationStat 2020; Wamala 2013; Worldpopulationreview Countries by Area 2020; Worldometers 2020; Worldpopulationreview 2020; Worldometers Uganda demographics 2020; Worldometers worldwide life expectancy 2020).

Lack of access to the internet is amplified by gender disparities. The Global Gender Gap Index (GGI) registers the gap between the social position of women and men with an aim of achieving gender sameness. A low GGI score is usually associated with developed countries, where commonly women and men are closer in social position than in the rest of the world. Uganda's index result is 65 (WEF 2020). When we look at an index designed to register the degree of discrimination faced by women—that is, to focus upon the extent to which women and men are differently treated—we see another story (Barnat et al. 2019). The Gender Inequality Index (GII) shows Australia, for example, to have few formal barriers to women's empowerment, while Ugandan women are very poorly protected in formal terms (rankings of 6 and 159 respectively; UN Development Program GII 2019).

Widespread poverty also restricts digital participation. The country's richest 10 per cent in 2017 owned over 35 per cent of Uganda's wealth (Mafabi and Oketch 2017; Kavanagh 2019) and despite some success in reducing the proportion of its population living in poverty, Uganda 'remains one of the poorest [countries] in the world' (Lubaale 2019). This situation is exacerbated by corruption (Transparency International 2017) and regionally situated ethnic-tribal and religious divisions (Cimardi 2015; Laruni 2015; The World Bank 2016: 10). Moreover, internet access has been restricted by government intervention through, for example, a controversial social media tax and censorship blocking some websites deemed pornographic or otherwise socially problematic.

Reports of government use of spyware and its collaboration with the Chinese technology firm, Huawei, have raised additional concerns regarding surveillance of citizens, opposition figures and government critics (Freedom House—Uganda 2019b).

In this context, internet penetration is relatively low, although growing, and usage is constrained by slow internet speeds and by limited access to electricity in rural areas, leading to a considerable digital divide. The extent of penetration has been variably reported at between 40 to 42 per cent in 2020 (Anonymous/PC Tech 2020; Freedom House—Uganda 2019b; Internet live stats—Uganda Internet Users 2016; Internetworldstats 2020; Kanaabi 2020; Kemp 2020c; Uganda Communications Commission Blog 2019). However, internet use in Uganda rose from .003 per cent of the population in 1995 to around 22 per cent in 2017 (World Bank 2017). Internet penetration rates have been increasing at speed mainly via the growing purchasing of mobile phones (Ybarra et al. 2006).

Most of those in Uganda who are able to access the internet for personal use, rather than solely through institutional settings like schools, are relatively economically privileged and educated young people located in the capital city, Kampala, (Wamala 2013). In other words, access to and use of the internet for romantic pursuits is not a widespread activity in Uganda. Rather, it is one which is more likely to be available to a particular narrow demographic. However, for this limited pool, it provides a private environment in which to receive individually tailored information and which can enable some freedom from social scrutiny—even though it is subject to stigma (Ferdman 2016).

Mingle2, the mostly free Ugandan version of a dating site/app makes use of a basic matching system where users can choose from random profile recommendations. An algorithm assists with matchmaking and the site offers ways in which to undertake regional searches in Uganda. This option likely is related to the marked racial/ethnic and religious diversity of the country. *Mingle2* makes a point of being 'easy, fast and effective'. All that is needed is to sign up and then find free matches 'in minutes'. The site boasts over 39 million members worldwide, presumably most from outside Uganda (*Mingle2* Uganda no date). A number of reviewers commenting on the site note that it has a high proportion of fake profiles (Dating Scout 2020c).

The analysis of *Mingle2* was augmented with offline conversational interviews as well as examination of dating profiles in local newspapers and television dating shows, in order to fully understand the potential for innovation in internet dating. In keeping with the demographic outline of internet users in Uganda, the great majority of profiles (18 out of 20) were younger than 35 years. All 20 profiles used English as the language of communication. Despite the considerable number of languages spoken within Uganda (Freedom House—Uganda 2019b), English is the medium of instruction at all educational levels and well accepted by the mainly elite internet daters. It thus provides a means to communicate across ethnic/language divides and allows access to potential partners beyond the geographical boundaries of Uganda.

This sample of Ugandan internet daters reiterate gender, race/ethnicity and (gendered age-based) heteronorms through contextually located understandings of racial/ethnic hierarchies, in which heterogamous potential partners are deemed especially desirable. For example, four of the ten women were very frank in communicating their desire to meet—white—partners outside Uganda. This inclination has been noted by a range of commentators (Sawlani 2012)—though several note this expressly to outline the dating advantages accruing to foreign heterosexual men who wish to date Ugandan women (Harris no date). While this interest in dating white/foreign men reiterates norms, it also allows the search prospects of Ugandans to extend beyond their immediate vicinity, a possibility which may be viewed as innovative within a country with strong ethnic awareness. The high valuation of whites/foreigners (muzungu) is not limited to Ugandan women and is not solely a question of a hierarchically organised and post-colonial society but also associated with traditions of hospitality (Beauvoisin 2015; Mercks 2015 YouTube).

Culturally specific intersections between age and gendered heteronormative assumptions regarding marriage and children were also plainly evident. Three women of 30 and over, were also candid in hoping their online profile would yield ‘a long term committed relationship’ and one stated she was ‘not here for Internet games’. Among the male profiles analysed, a 49 and 56-year-old were ‘looking for serious women to marry’. By contrast, the other male profiles and two of the female profiles in their early 20s were more casual, even coy in their presentation of themselves. Social maturity for men and women is aligned with marriage in Uganda (Hendricks 2018), and the older one gets without settling down and getting married the more elusive this maturity (Nannyonga-Tamusuza 2005: 134–5). For women, regardless of their marital status, the passage to womanhood is aligned with giving birth. In the setting of the high value placed on women having children, it is perhaps unsurprising that ‘Uganda’s fertility rate has been among the highest in the world for 40 years’ (PAI 2010: 2). The 2010 report by PAI noted that few women felt they could make decisions about their healthcare and men commonly considered that husbands should be primary decision-makers regarding how many children to have.

The women on *Mingle2* expressed a keen interest to have children, and although not stated directly, this was more urgent than long-term commitment. Having children surpasses the urgency of marrying for women because, a woman who has never given birth no matter their age is seen as ‘lazy . . . useless’ (Nannyonga-Tamusuza 2009: 375).[6] The question whether those putting forward profiles have children or not is especially sensitive for women, for whom the ‘useless’ label shadows their status as proper women in Ugandan society. This then adds, especially for women over 30, compelling priority to their desire to find men who can be intimate/sexual partners.

However, going online is not always seen as the best way to find partners and these women are disrupting gender norms around what is considered proper.

> [P]utting oneself out there through the newspapers or television is somewhat less embarrassing, than going online. Women who go online they are thought to be easy.
>
> *(Fiona, 33)*

> Girls who go online, are most likely to come from upper middle class educated backgrounds, one would not expect them to struggle finding someone, but those who go to the newspapers have nothing to lose and everything to gain, they are not from the same social background as those going online, even though 5 or so years ago, it was also demeaning to put one's image in the lonely newspaper columns.
>
> *(Paul, 35 and Fiona, 33)*

Despite the uncertainties of going online and the risk of being considered 'easy', this more unconventional path allows women to initiate contact with men (see Chapter 2).

> [G]irls in the past have not been the aggressor in the pursuit. Technology helps them to be able to approach . . . even the shy ones. Those who go on *Bukedde*[7] and *Abanonya*[8], majority of them are women.
>
> *(Tendo, 72)*

In the Ugandan interviews, a woman taking the initiative in dating is described as taking up different and unconventional behaviour within the national context. Online dating allows women in Uganda to engage with what Judy Wajcman describes as a technofeminist approach (2004, 2007), enabling increasing opportunities for fluidity and flexibility as against what may be negotiated regarding finding potential local homogamous partners offline. Though Ugandans face social stigma associated with putting themselves online, use of closed Facebook pages, and other social media platforms such as *WhatsApp* groups is escalating, along with the growing use of internet dating (NBS 2016; Salmon 2020).

In sum, online dating appears to provide relatively privileged Ugandan women with opportunities to have some control over dating, in ways that recall our earlier chapters on shifts in the behaviour of UK and Australian women. Nonetheless, the profiles and interviews in the Ugandan study reveal that these women also might use the privacy and autonomy provided online to undertake reiterations of internalised racialised/ethnicised norms of desirability. It is possible that these norms of desirability reflect the attractions of white/foreign men's (and typically older white/foreign men's) probable greater financial resources and professional prospects. Many studies worldwide report an inclination on the part of women to prefer men with such resources and prospects (though this is decreasing as women become more economically independent) (see Chapters 3 and 5).

However, it may be that the seeming desirability of white/foreign men reflected in these profiles and interviews with Ugandan women is associated with them wanting to find 'exotic' strangers whom they deem to be less traditional, less patriarchal, than local men (Kukunda 2016; Wyrod 2009). This is the obverse of the more widely known spectacle of white male tourists seeking supposedly 'traditional women' overseas (Berg et al. 2020; Tepanon 2006). Ugandan women in this and other research regularly complain that their Ugandan male suitors are unreliable and unfaithful, citing such men as failing to mention that they are already married or cohabiting, have other girlfriends and/or have children (see for example, muzungubloguganda 2014) In other words, despite the racist/ethnic overlay of Ugandan women's valuation of white/foreign men, there are hints here of these women attempting to break away from proximate male dominance. And finally, the profile materials and interviews indicate that Ugandan women and 'older' men are embracing technological change to meet conventional social norms around heterosexuality, hetero-coupledom and gender with regard to getting married and having children, while simultaneously evading some familial, ethnic and social restrictions around arranging to meet, partnering and partner acceptability (Girls not Brides no date; Hague et al. 2011; OHCHR no date).

Victoria Milan, Denmark

Denmark[9] is a small, affluent country with a relatively small population, which is noticeably homogeneous in terms of socio-cultural characteristics.[10] Indeed, what has been described as 'cultural levelling' has been a feature of Danish society in the twentieth century, linked with expansion of the Scandinavian welfare state, 'resulting in relatively small socioeconomic and educational differences among the Danes', and linguistic and religious standardisation (Madson 2013: 117; Erasmus 2017; Iversen 2006: 75–76). This remarkably strong coherence in Danish society is notable in being associated both with a welfare egalitarianism—that is, with practical, pragmatic collectivism—and with a distrust of collectivity in favour of self-sufficiency, of 'standing on one's own two feet', in relation to other social arenas. While there is a determined support for citizens' obligation to maintain a welfare state which acts as a symbol for Danish social solidarity, personal autonomy is also valued, including independence from family (Verdenpressen 2017; Iversen 2006: 76–78).

Denmark demonstrates a very robust concern with social equality in terms of economic/welfare and gender aspects of social equality, though this concern is not unlimited (Holtug 2013; European Commission on Denmark 2020: 4; Causa et al. 2016: 7). Gender equality is part of Danish cultural specificity and is in practice high in the region (Soergel 2020). In 2019, Denmark topped the SDG Gender Index scores worldwide (Equal Measures2030 2019).[11] There is broad support, for example, for the ideal of equal pay and Danes are confident of their government's actions in relation to gender equality (Andersen and Shamshiri-Petersen 2016: 5, 8; 10, 14). Thus, when the Danish Ministry of Culture conducted a survey titled the 'Danish Canon' in 2016, the Ministry found that the ten most important values and traditions were seen by Danes as including the Welfare Society, Freedom and Broad-Mindedness, as well as Equality before the Law and Gender Equality (Verdenpressen 2017).

Yet, in 2020 Denmark was the only Scandinavian country to fall out of the top ten worldwide for gender equality (WEF GGI Index 2020). Moreover, some scholars and commentators assert that, for instance, Danes are significantly less committed to challenging traditional gender hierarchies than their Nordic neighbours, view feminism as a 'dirty word', and have barely come to grips with sexual harassment and assault with little concern about the claims of #MeToo (Andersen and Shamshiri-Petersen 2016: 5–6, 10; Dellisanti 2017; Orange and Duncan 2019; Soergel 2020). Furthermore, the term 'equality' applies for the most part to those who are deemed 'insiders' within the social rubric of Danish society. Racial/ethnic and to some degree religious equality, for instance, are not featured. In other words, there is a downside to Denmark's marked social solidarity. The cosy contentment that Danes describe as 'hygge' and the nation's glowing status as a country of equality, open-mindedness and happiness (MakingItMagazine 2013; Savage 2019) appear in this light as largely restricted to the familiar, to those who are the same.

Denmark's apparently progressive reputation seems derived from a deep-seated insider/outsider distinction involving discomfort with difference and diversity. In this context, Hedetoft asserts that '"Danish multiculturalism" is an oxymoronic notion,' and that cultural diversity is frowned upon as alien, as a foreign notion (2010: 11; Lægaard 2013; see also Østergård 2012, 2000; Selmer and Lauring 2013: 9; Tawat 2018). Furthermore, some 'outsiders' may be even less acceptable than others. Very frequently, debates about immigration are fixated upon Muslims who presently make up the largest non-western grouping immigrating to Denmark (Freedom House—Denmark 2020b; Holtug 2013: 191). In brief, gender equality in Denmark is not perhaps as secure or actual as might be assumed and widely accepted ethnocentrism demonstrates ongoing discomfort with cultural diversity, providing two instances of seemingly paradoxical features of Danish society. In these features may be found both unexpectedly regressive and heterodoxical possibilities for internet dating.

Internet dating in Denmark occurs within the context of very high access to and usage of the internet, as well as high levels of skill in its population. In 2017, it ranked just above the UK IDI at a score of fourth in the world (ITU-IDI 2017). There is little in the way to restrict internet usage. Indeed, Denmark scored 97 out of 100 in world rankings of internet freedom (Freedom House Country Data 2018; Shahbaz 2018;). Internet penetration stood at 98 per cent in January 2020 (Kemp 2020d). Danish society, especially public administration, is highly digitised (Denmark 2020: Ecommerce Country Report 2020;

Træen et al. 2018). It is no wonder that Danish social links are also singularly digitised and show a significant level of utilisation. For example, almost half of Danish single people under 35 used online dating services in September 2018 (Paisley 2018).

Internet dating is widely viewed as an acceptable way of meeting others by younger and older Danes (Jørgensen 2011; König et al. 2018; Træen et al. 2018). The context for this is associated with life expectancy and availability (Terry 2019; Worldometers Life Expectancy by Country 2020). This, along with one of the highest divorce rates in Western Europe (in 2019 around 35%), provides a potential context for widespread online dating amongst older people (Henley 2019).

However, Denmark is also reported as having high levels of marital infidelity, arguably having the most unfaithful partners in the world, with Danish men being significantly more likely to be unfaithful than women (Mellish 2015; Seferou 2017; Statista 2020b;). This information may well have some implications for our examination of a particular Danish internet dating site. While Denmark is strongly associated with gender equality (Matveyeva 2020; Mellish 2015; Seferou 2019; VirtualWayfarer 2015), as we have pointed out equality between men and women in forming intimate relationships is not apparently entirely given or straightforward. Hence, we have focussed upon a dating site that would probably be viewed as at odds with progressive directions in gendered heteronormativity, let alone a source of such directions.

Victoria Milan is an international dating site that specialises in connecting already partnered people who wish to have an affair without the knowledge of their partner. Its focus thus appears out of step with the Nordic/Scandinavian regional approach and with the widely recognised and much celebrated national acceptance of gender equality in Denmark. Norwegian entrepreneur Sigurd Vedal founded the *Victoria Milan* site in 2010. It offers approximately 30 different country versions. The Danish site provides notable features, many associated with its focus on secret encounters, which entail some dissonance from the Danish public stance regarding relations between men and women. For example, facial features on photos can be blurred or a sticker like a mask put on them, a 'panic button' enables users to press a button which takes them to a common website like YouTube so that the user's partner 'is none the wiser', virtual gifts can be sent and received, and unsurprisingly there is a heavy emphasis on user confidentiality and the ability to delete the site (Navarro 2015; *Victoria Milan* Review 2020). While the site's assertions regarding confidentiality include enabling members to be assured that its name will not appear on their billing statements, the option concerning deleting the site may not necessarily be as secure as *Victoria Milan* claims. However, the site has upscaled this possibility following hackers successfully accessing *Ashley Madison*, a site with a similar focus on secrecy and infidelity.

These features challenge Denmark's innovative public stance in global terms concerning a national commitment to equitable gendered heterosexual relations. Indeed, *Victoria Milan*, appears to offer regressive options. Denmark, in short, can provide surprising reversals of what might be expected. In our analysis of other country-based sites and profiles public norms of gender and sexuality are frequently undermined by intimate private practices in dating. Yet, in the case of Denmark we see instead a regression in these practices on the selected dating site from comparatively more equitable public norms. Most often public norms are more conformist than the micro-politics of personal interactions, but not always it would seem. Denmark, rather than simply offering an idyll of progressive innovations, might instead be a reminder of the deeply uneven character of social change.

The first 40 profiles (20 men and 20 women) were selected from a wide-ranging search. In order to access profiles on the site it is necessary to register and create a profile. For this study, a 'basic' profile was created free of charge, which allows browsing of other people's profiles but not interaction with these potential dates. Following creation of this, a search was performed for men within 10 km of Copenhagen, aged between 18 and 50, whose profiles included pictures. This produced well over 1000 hits. A similar search for women produced approximately 800 hits. For both men and women, most profiles belonged to users in their 30s and 40s. No users were aged under 20, and there were significantly fewer in the 20s and 50s age brackets. The majority of users identify as white Danes.

Their profile layouts are somewhat different from many other heterosexual dating sites. Clicking on a profile leads to an in-depth profile page with a large image and three text-based sections. Each profile page has a 'headline' giving user name, age and location, with a sub-headline summarising the user, for example, whether they are married or not, what kind of relationship they are looking for (for example, online flirtation or full-on relationship) and their reason for being on the site (such as feeling neglected by partner). The first of the text-based sections is titled 'About me' and comprises 24 categories focussed on physical appearance/lifestyle, including height, weight, and unlike many other dating sites, 'sex drive'. The latter signals the rather more bluntly sexual orientation of the site. There are a range of other hints of this. For instance, in the second section members of the site are asked to summarise themselves with questions like 'I am turned on by . . .' and indications of body types and sexual practices. The third section titled 'My Ideal Match', also asks about 'sex drive' and sexual orientation. The design of the site carefully controls the profiles by relying heavily on lists of predefined categories.

Responses in the 'About Me' section are revealing in terms of gender differences. Men's profiles listed qualities like (starting with most popular) good sense of humour, flirty, active, empathetic, sporty, honest and confident. In contrast, in women's profiles, the most popular responses (starting from the most popular) were, honest, passionate, open, good sense of humour, spontaneous, empathetic. Whilst some overlap can be seen, there is a clear distinction between what is considered attractive for men compared to women. For example, the men's profiles, similar to the UK profiles, show a much stronger emphasis on physical activity and sport, combining an agentic energy with claims to emotional sensibility. There is little here to distinguish the mode of gender differentiation in the Denmark profiles with those from UK instances.

Victoria Milan, like the British sites *PoF* and *Match.com, UK*, reproduces certain norms around bodies and genders through the 'tick-box' requirements of selections. This tick-box format means sticking to 'known pathways' such that daters can 'rarely venture into the unknown' (Garde-Hansen and Gorton 2013: 7). The site's reinforcement of 'known pathways' undermines, at the micro-level of intimate interactions, Denmark's remarkable collection of indices indicating its adherence to social change through advancing gender equality.

The second notable feature of *Victoria Milan* profiles is its reframing of heterosexual non-monogamy, but in ways which may trouble assumptions regarding the importance of honesty in achieving gender equality. While, as we have outlined earlier (see Chapter 4 in particular but also 3 and 6), internet dating sites are precisely set up for simultaneous multiple connections between daters to arise, this facilitation of what we have described as 'mundane polyamory' is capable of enabling unconventional directions regarding gender and heterosexuality. Mundane polyamory unsettles gendered heteronormative assumptions regarding flirtation, romance, sex and coupledom in that it does not take as given that there is only one soulmate for each person, that one person is all that is needed to meet another's needs, and that the one romantic/sexual mode of intimacy must always be the goal. It may also encourage women daters to initiate and take charge of dating practices while potentially disturbing notions of masculine authority and control. However, there is potential dissonance between public emphasis on honesty in both men and women's self-presentations on their profiles as key to potentially egalitarian relations and their 'private' engagement in a dating site devoted to the practice of deliberate dishonesty.

Sites such as *Victoria Milan* cannot be viewed as simply using the technical affordances of internet dating to facilitate extramarital relationships but rather are directly engaged in redirecting potential unconventionality back into recuperation of traditional hierarchical norms concerning heterosexual gender relations. The *Victoria Milan* site reinstates normative conceptions, for example, of masculinity as founded in pursuing, in predatory interactions with women, in cheating women in order to get access to sex, in a reframed resurrection of 'the mistress' such that women are implicated in men's interests, and in an entitled upholding of men's 'freedom' as arising from dishonesty in relation to women. *Victoria Milan*, through its open advertising campaigns, seeks to decrease the stigma associated with this form of non-consensual heterosexual non-monogamy via the technical possibilities of the digital medium which make it easy for customers to access the site services (Baym 2010), and by casting the dating practice it offers as liberating, innovative and an 'equal opportunity' way for both men *and* women to respond to lack of attention or loss of passion in marriage or committed relationships—even providing a means to support the existing marriage/commitment (Dating Scout 2020a; *Victoria Milan* Blog no date; *Victoria Milan* no date). Yet *Victoria Milan*'s supposed contribution to social 'innovation' is at the very least deeply equivocal in terms of the Danish concern with gender equality. It is important to note that non-monogamy is not all the same in terms of social change. Indeed, the non-consensual form of heterosexual non-monogamy the site offers closes down rather than opening up challenges to hierarchical heterosexual gender norms. The preponderance of men on the site is indicative. *Victoria Milan*'s membership is vastly unbalanced in terms of men and women users: of its 235,000 users in Denmark it is reported that only 20,000 are women. In a related vein, when *Ashley Madison*—a similar international site for infidelity—was hacked in July 2015 (as noted earlier), according to the group of hackers (the Impact Team) 90 per cent of its supposed female users were fembots (Navarro 2015).[12] It seems likely that this is a feature of *Victoria Milan* today (Open Eye 2019).

RSVP, Australia

Australia's stability as a democratic nation and its long-standing prosperity have provided and continue to provide a sound basis for progressive change enhancing political and civil prerogatives (Australian Human Rights Commission 2017, 2013; Evans et al. 2020: 1; Freedom House—Australia 2020a; Williams 1999). Its 'constrictive' population pyramid, like those of Britain and Denmark, is characterised by a lower proportion of the population being made up of younger people (declining birth rates) and longer life expectancies (Worldometers Australia Demographics 2020; Worldometers worldwide life expectancy 2020). In 2020, Australia's population of around 25.5 million is very largely urban—about 86 per cent (Worldpopulationreview Australia population 2020; Worldometers Australia demographics 2020). While it is the world's sixth largest nation and is one of the least densely populated overall, its population is highly concentrated (Worldpopulationreview Countries by Area 2020; Worldpopulationreview Population by Country 2020).

Along with this relatively comfortable economic position and a modern urban lifestyle, Australia has a long history of commitment to anti-authoritarian egalitarianism associated with the powerful significance of a national mythology upholding 'a fair go' for all (Barry 2017; Chipperfield 2019; Sheppard and Biddle 2015). This history underwrites support for a welfare state and a public health system. Although compared with many other liberal democracies including Denmark, Australian citizens are more inclined to distrust government authority overall and there is a national narrative of disregard for authority in favour of informality and egalitarian 'mateship' (comradery; Evans et al. 2020: 2, 6; Evason 2016), but they are nevertheless likely to accept government directives (Hirst 2004; VicHealth 2003).

While Australia registers public support for robust, socially liberal values—including in relation to an increasing embrace of plural ethnicities (Markus 2019; Moran 2011; Rajadurai 2018)[13] and gender equality (Evans et al. 2020: 17; World Values Survey Wave 7 2017–2020)—this sits alongside some regressive views and practices (Beck et al. 2019). Elston observes in this context that Australian national identity is necessarily 'complex and fractured' with long-standing contradictory directions increasingly in tension (Beck et al. 2019; see also Crossley 2018; Holmes and Pinto 2013; Australian Human Rights Commission Section 7 2011; Jotanovic 2019; Larmour 1999).

Just as Denmark has a national agenda strongly framed in terms of gender equality, so Australia has long prided itself on its diverse civic polity (Mason 2019). The 2019 Mapping Social Cohesion survey showed a high degree of ongoing public acceptance of multiculturalism, with 85 per cent agreeing that it had been good for the country (Markus 2019; Soutphommasane 2016). Nevertheless, Australia's history is deeply marked by its foundational violent colonial dispossession of the Indigenous population and a racial exclusion immigration policy (the 'White Australia' policy) begun in 1901 and progressively dismantled by 1973 (Jones 2017; Rajadurai 2018). Racially inflected narratives still bearing the lingering marks of the White Australia policy continue to haunt Australia's treatment of (predominantly non-white) refugees and asylum seekers (Bolger 2016; Jones 2017; Jupp 2007; Koleth 2010; McMaster 2001; Neumann and Tavan 2013: 1–8; Phillips 2015).

We note also the dissonance between Australia's increasing contemporary expectations and requirements regarding gender equality, and a long-standing masculinism deeply embedded in Australian society and its national story (Cousins 2020; Hirsch 2018; Marti 2018). While in terms of the 2020 Global Gender Gap Report (GGI) Australia scores highly on women's educational attainment and health, it continues to lag in terms of women's economic participation and political representation. Indeed, Australia has slipped from 15th out of 115 countries in 2006 to 44th out of 153 (Dent 2019; PerCapita 2020; WEF 2020). In short, though in the more comprehensive SDG Gender Index ranking (Equal Measures2030 2019) Australia scores tenth in the world, the difference between the position of Australian women and men has widened. This reflects entrenched structures of gender disparity—such as a continuing wage gap that has barely changed over two decades (Workplace Gender Equality Agency 2020), and deeply ingrained social attitudes (Evans et al. 2018).

In the light of these aspects of the Australian context, we will now turn to use of the internet. Australia has a high level of internet penetration at 88 per cent in 2020. Internet access is affordable for most Australians (Freedom House 2019a; Granwal 2019; Kemp 2020a; ITU-IDI 2017) and there is little to constitute limitations on access to and usage of internet dating. Our focus in this Australian study involves 40 dating profiles on a large Australian internet dating site, *RSVP*, which was deliberately chosen because its membership pool is older than is usual (Dating Scout 2020b; Hockey 2020; Sheftalovich and

Smith 2020a, 2020b). *RSVP* profiles are set up with a profile photo and short snappy lead as the starting point, followed by basic information about the person which is coupled with how the person sees their ideal partner in terms of this information. For example, the person's age is coupled with the preferred age range of an imagined ideal partner. This is followed by brief statements regarding what the person is looking for in a partner and regarding their own interests in music, books, sports and other entertainment activities. Finally, members are offered a short 'personality' test which can be made available to those perusing their profiles. Like the other sites we have examined, *RSVP* members who wish to contact others are required to pay a fee. Most profiles from *RSVP* belonged to users in their forties, with much fewer under or over this age range. An even number of men and women's profiles were chosen from a selection of the first 20 profiles (two in their 20s, three in their 30s, 10 in their 40s, one in their 50s, three in their 60s and one in their 70s). Educational backgrounds were roughly divided into half and half in terms of tertiary education. These *RSVP* members did not mention their ethnic/racial identities, though superficially all bar one appeared to be white.

While recently several internet dating sites/apps (though by no means all) have eschewed racial/ethnic filters (Thomson et al. 2020), on *RSVP* it is possible to describe one's own 'ethnicity/background' by choosing from 11 possible categories and selecting from these same categories in choosing 'my ideal partner'. As noted earlier (and in Chapter 4), heterogamy in Australia is not uncommon. Of those who were in couples prior to 2011, around 1 in 4 relationships were inter-ethnic (Lass 2019). Watts reports that by 2016 over a third of marriages in Australia were between people from different countries of birth (Watts 2019: 104). Inter-ethnic partnering is associated with waves of migration to Australia with increasing numbers of such couples across all ancestry groups but significant increases amongst those of Greek, Lebanese or Chinese backgrounds. In 2013, 60 per cent of third-generation migrants from these groups are reported as forming inter-partnerships (ABS 2013; Khoo et al. 2009; Lass 2019; SBS 2013a).

Yet, online dating also shows evidence of resisting this trend to maintain cultural/religious and other community bonds. This is perhaps particularly important in a multi-ethnic nation which shows little antagonism to inter-ethnic partnership. In this context, a range of online dating sites offer culturally specific matchmaking services including Greek, Aboriginal, Korean, Jewish, Indian and Muslim services (Giakoumelos 2017). Although *RSVP*'s survey in 2012 found that two thirds of respondents did not rate ethnic background as important when identifying a potential partner, nonetheless around a third did (SBS 2013a; 2013b). This largely appears to reflect positively choosing to maintain cultural distinctiveness and identification. Moreover, ethnic dating sites are likely to incorporate and embrace the significance of family decision-making (which may include the use of professional matchmakers) in finding suitable partners, as well as providing a response to the problem of small dating pools.

These dating sites may also attend to gendered pressures to marry and have children by a certain age, as we noted in relation to Uganda (Dwyer 2016).

Overall, the most evident feature of these self-presenting profiles is similar to that found in the UK and Denmark studies. Women in the Australian sample are more likely to mention their upbeat fun disposition and sociability, while men regularly mention their level of activity and fitness and relaxed temperament. Male daters write markedly less than women, thus providing far less information about themselves. The Australian men, like the men on UK sites, are inclined to provide far fewer professional photographs, with seemingly little attempt to appear more desirable. The gendered differences in orienting vocabularies and visual presentation employed in the profiles are indicative of intransigent patterns around heterosexual gender relations. For example, heterosexual men are likely in these profile pictures to locate themselves in relation to markers of traditional gender roles such as providing photos in which they are presented with their cars, motorcycles, or boats. Such tropes might seem to suggest that heterosexual men, and perhaps particularly older heterosexual men over 40, are attempting to uphold/recuperate quite traditional conceptions of masculinity. Relatedly, in investigating Australian dating sites like *RSVP*, there is noticeable gender differences around what ages are deemed desirable that can be observed in the ideal partner age range listed on profiles. Heterosexual male daters prefer, in principle at least, younger women than themselves, and very frequently much younger. Heterosexual women daters on RSVP are looking for men of around their own age.

However, despite the reproduction of traditional heterosexual gender roles in the *content* of the profiles, there are indications of innovative *behaviours* that arise from dating-site design and how the profiles are organised. One evident practice that we have outlined in this and earlier chapters is what we have termed mundane polyamory. Dating several people at once is a frequent, indeed rather usual, and comparatively easy endeavour in online dating. The profiles themselves and mechanisms for responding to them are devised to enable daters to explore a range of profiles and thus engage with more than one potential dating prospect at a time. However—in the context of the reiteration of traditional gender roles evident in the Australian profiles—it is intriguing that *RSVP* research on its membership (RSVP Date of the Nation Report 2015) and an associated post (RSVP team 2016) reveal that this innovative challenge to monogamy does not apparently follow gender stereotypes. While 54 per cent of heterosexual men surveyed accepted mundane polyamory, intriguingly the research showed that women daters are decidedly more willing to engage with this innovation: '63% of women say it is ok in the early stages of dating to continue dating others' (RSVP Date of the Nation Report 2015; Beasley and Holmes 2016: 99; see also Patel 2018).[14] In this context, an Australian woman who particularly focussed on using two dating sites (*OkCupid* and *RSVP*), advocates trying out several sites, narrowing down to ones that are personally most suitable and then using those sites to meet people whose paths you might never cross—that is, to meet people 'from outside your bubble' *(Ina, Australia*—cited in Sheftalovich and Smith 2020b*)*. For Ina and many other women, online dating is precisely to be recommended because of its multiple possibilities.

Within internet dating both men and women can consider a variety of possible partners, and actively undertake a range of decision-making practices. The shared knowledge by users that this is the case has the potential to counter traditional conceptions of masculinity as active, in control and/or sexually predatory. As noted earlier, men's profiles on the Australian site, *RSVP*, regularly, indeed almost invariably, highlight how 'easy-going', 'relaxed' or 'laid-back' they are. This reiteration may well be a disguised language for acknowledging women's oft-stated concerns about men bossing them around. Whereas once these men might have responded negatively to heterosexual women initiating dating contacts and dating multiple partners, online many are willing to accept this shift in gender and sexual relations.

As we observed in relation to the UK profiles, it is possible that the 'curating' of gendered public self-presentation (see Chamourian 2017) may be one of the most averse to shifts towards gender equality in the field of internet dating. Oddly enough, sexuality, sexual practices and acts—often seen as one of the more hotly contested arenas of personal and social life in terms of change (Fahs et al. 2020; Ford Foundation 2005)—may prove more amenable to flexibility. This point confirms that social change is highly uneven not only in the sense of cultural diversity, but additionally even within a particular sphere of activity such as internet dating, and perhaps especially in the micro-politics of such a sphere. Is there evidence here of both the comparative difficulty of changing public narratives as against private practices and of the comparative intransigence of gender performance as against sexuality performance? Are such considerations capable of generalisation beyond internet dating? And, finally, do these considerations say something consequential about specific locations which might enable social change in relation to gendered heteronormativity?

Schematic extension study—materials and methods: Argentina, China and Turkey

This schematic study involves a restricted extension of materials and methods. It extends the book's main focus on internet daters in the UK and Australia to provide some limited rebalancing away from the anglophone developed world towards a wider range of cultural contexts. The schematic study also extends our collaborative pilot study of internet dating sites and profiles from four countries by a further three, and in the process offers a lens on sites and profiles across five of the seven continents across the world.[15]

However, it is important to note that this schematic extension of the collaborative pilot study of internet dating sites and profiles involves a considerably less developed line of enquiry and requires further work. Rather than enabling us to outline what we have described as 'snapshots' in relation to the pilot study, the schematic study generates broad-brush observations. In this relatively embryonic research, we adopted another methodological approach which is different from either the detailed focus on UK and Australian internet daters or the selective examination on dating sites and profiles in the UK, Denmark, Uganda and Australia. Instead, we chose one dating site in each of the three countries (Argentina, China and Turkey)—that is *Latin American Cupid*, *China Love Cupid* and *LoveHabibi* respectively—in this schematic study. The sites were chosen on the basis that English was or could be used and/or that English translations were reasonably easy to undertake. When loading each of these three dating sites, we selected the first five profiles of men and then the first five of women to appear between the ages of 25 and 65, although sites sometimes had specifications that did not allow a full choice of age range. Commentary on the 30 profiles involves general observations about notable national characteristics associated with the sites and profiles. Rather than discussing or quoting individual profiles, we anonymised individual materials by providing amalgamated accounts of common self-presentations, while also providing pseudonyms to retain a sense of the personalised nature of these self-presentations.

In this schematic study we employed the same form of content analysis to analyse textual materials, including visual texts, as we did in relation to the pilot study (Cavanagh 1997). Context is of significance in this schematic research as it was in the collaborative pilot study given significant national/cultural variations in internet access, in the extent of use of internet dating and the degree of acceptance of that use as a means to meet potential partners. While in the pilot study the UK, Denmark and Australia share high levels of internet access, dating site use and acceptance of this use, this is not the case with regard to Uganda. Differing access, usage and acceptance impact in different cultural contexts upon potential for social innovation within internet dating. When we extend the number of countries under study this becomes even more evident.

Broad-brush observations arising from the schematic study

In this schematic study of Argentina, China and Turkey, we provide perspectives which expand the field of analysis we consider, opening up directions that are vastly different in a range of ways. The populations of these countries, alongside those in our main study, vary from about 6 million to the most populous in the world (Denmark 5.7 million to China over 1.4 billion) (Worldpopulationreview 2020). In relation to one of the most widely employed indices to assess gender relations we see a similarly disparate array in relation to Global Gender Gap Index (GGI) scores (Denmark 14, UK 21, Argentina 30,

Australia 44, Uganda 65, China 106 and Turkey 130; WEF 2020). There are intriguing further points to make regarding the distribution associated with the GGI Index. While Africa and the Middle East amongst other regions are markedly gender unequal overall, Uganda is somewhat more gender equal than many of its African neighbours and Turkey is similarly somewhat more gender equal than its Middle Eastern neighbours. By contrast, Australia is a relatively poor performer for an affluent developed nation when considered alongside comparable English-speaking countries (like Canada, the UK and New Zealand) and Western Europe including Scandinavia (WEF Performance by Region 2020).

IDI rankings comparing access, usage and skill levels in relation to internet usage indicate that only three (the UK, Denmark and Australia) of the seven countries examined have high levels of internet utilisation capacity but this technical data tells us little about the extent to which internet dating in different cultural contexts may enable unconventional directions in gendered heteronormativity. To explore this issue, we broadly consider the collaborative pilot and schematic study of internet dating profiles in the light of the World Values Survey (WVS) 1981–2015 (World Values Survey Wave 7 2017–2020). The WVS uses Inglehart–Welzel cultural parameters to show in generalised terms how major cultural zones, and countries within them, shift in values over time. For example, there are nine major cultural zones identified (including African-Islamic, Latin America, English Speaking, South Asia, Orthodox, Catholic Europe, Protestant Europe, Baltic and Confucian). Hence, Uganda and Turkey are located in the African-Islamic cultural zone, Britain and Australia in the English-Speaking, and Argentina and China in Latin American and Confucian zones respectively.

There are also two main axes upon which these cultural zones, and countries within them, are located: these are the vertical axis of Traditional to Secular-rational values and the horizontal axis of Survival to Self-expression values. The first axis represents the movement in values from societies which emphasise 'religion, parent-child ties, deference to authority and traditional family values', toward societies which express 'opposite preferences . . . with less emphasis upon religion, traditional family values and authority'. The second axis represents the movement in values from societies focussed upon material security towards societies that prioritise civil and political expression and equality for citizens including minorities. Some countries change slowly while others experience more rapid shifts in values as well as uneven shifts in values, and the speed of change in each country may alter over time.

We include two versions of the WVS Cultural Map, one of which is older but includes Uganda (2005–2008) and the other of which is the most recent available and does not label Uganda.

The WVS has been criticised on several grounds, including that national borders are not a decisive means to delimit cultural settings (Matai and Abrudan 2017) and the basis for determining the cultural zones in the WVS is

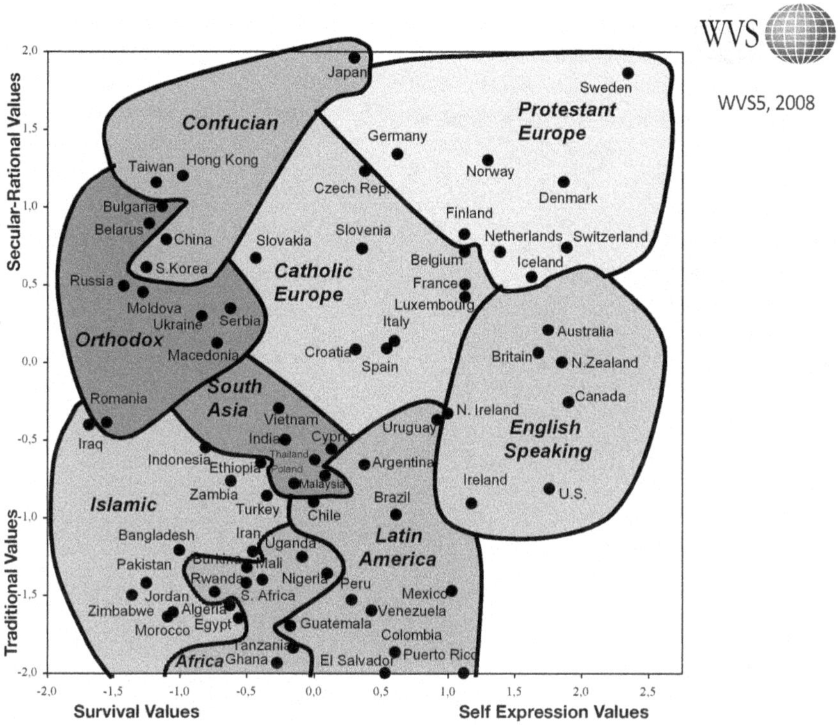

Figure 7.1 The World Value Survey Cultural Map 2005–2008. A colour version of this figure is downloadable from: www.routledge.com/9780415720694.

Source: R. Inglehart and C. Welzel (2010) 'Changing Mass Priorities: The Link between Modernization and Democracy.' *Perspectives on Politics* 8(2), June: 554.

varied (including being constituted by region, religion, political organisation and language). Perhaps more importantly, the WVS use of Inglehart–Welzel cultural parameters rests upon a form of modernisation theorising about social change. This revised form elaborates on the argument that economic growth is associated with the flourishing of democracy and individual personal expression, by adding the impact of cultural zones. Modernisation theories have been subject to considerable debate regarding their linearity, determinism and inclination to assume that Western values are superior and represent the overall direction of historical change (Acemoglu et al. 2009; Beasley 2017a; Bernstein 2007; Saito 2007; Welzel et al. 2003). Relatedly, the two axes for determining cultural parameters and change over time may be disputed. For example, China is placed as strongly Survival-oriented on the horizontal axis and as strongly Secular–rationalist on the vertical one. While the first assessment

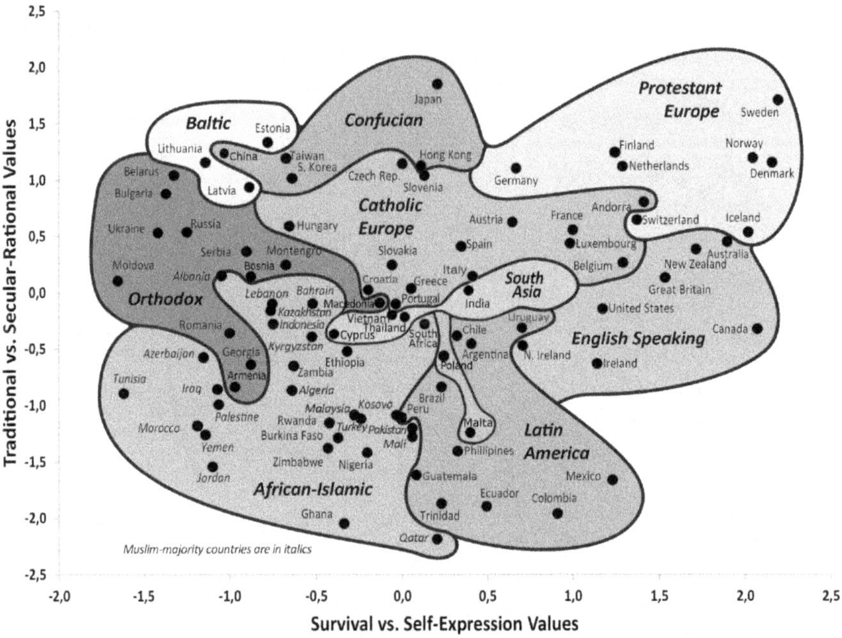

Figure 7.2 The World Value Survey Cultural Map (WVS wave 6) 2010–2014. A colour version of this figure is downloadable from: www.routledge.com/9780415720694.

Source: R. Inglehart, C. Haerpfer, A. Moreno, C. Welzel, K. Kizilova, J. Diez-Medrano, M. Lagos, P. Norris, E. Ponarin and B. Puraneneds (2014) 'World Values Survey: Round Six—Country-Pooled Datafile', [https://www.worldvaluessurvey.org/WVSDocumentation WV6.jsp. Madrid: JD Systems Institute].

seems unremarkable, the second seems to us at odds with the pronounced and widely recognised emphasis on parent–child links, family bonds and deference to authority in scholarship and our sample on China (Blair and Madigan 2019; Wong 2014). Apart from markedly lesser reliance on religious values and a degree of individualisation (Yan 2010), on every other parameter China appears to be relatively 'traditional'. Autonomy and freedom from tradition, so far as these are publicly acknowledged, often appear more of an ambition than a reality in terms of intimacy (Wang and Nehring 2013).

Nevertheless, despite such reservations, the WVS offers a means to consider the countries we examine in terms of innovations in heterosexual gender relations. It enables a viewpoint on possibilities for change in different countries.

Argentina

In Argentina, there are prominent constraints upon unconventionality with regard to daters' investment in the Catholic faith, close engagement with family and the critical significance of weddings, marriage and children (Hendricks 2018). Moreover, gendered subjectivities remain shaped by gender polarity and inequality linked with machismo. Indeed, such is the centrality and dominance of men that hierarchical gender relations often trump religious and familial morality (Carnes 2017; Castro 2016; Farah 2015; Goñi 2016; Timerman 2018). In internet dating, men's infidelity is reported as widespread and women's self-presentation is often marked by sexy and expressive bodily display. Tina, a young woman in her early thirties, living in Buenos Aires with her family and her 10-year-old son, describes herself as 'very attractive'—a not uncommon description for women on *Latin American Cupid.* On her profile she is wearing a revealing low-cut blouse and states that she is seeking a 'gentleman' who is financially solvent, will know what she needs and will take care of her. This amalgamated profile indicates a style of embodied gendered performance and self-presentation which certainly can be seen on the UK, Denmark and Australian dating sites but is considerably less common. Body image for women is rigorously policed in Argentina. Indeed, Argentinian national health insurance pays for one free plastic surgery operation each year (Maldonado-Salcedo 2017). Presenting a sexy image is featured in many women's profiles we surveyed as well as those we selected for closer examination. This was not evident in the men's profiles.

As might be expected from its middling positioning on the WVS Cultural map, Argentina has a mixed national narrative when it comes to gendered heteronormativity. Although its values regarding heterosexual gender relations are in many ways traditional, it also has a long history of female political representation. Its record of female leadership is second in South America and outranks OECD countries like Denmark (Bruggemann and Pence 2019). Argentina also shows increasing signs of shifts towards liberalising social policies which challenge gender inequality even if the realities of cultural norms are evolving more slowly (Conn 2018; Lopreite 2013; Hendricks 2018; Prince no date). The Argentinian profiles show the continuing weight of strongly gender differentiated self-presentations which, for instance, require women like Tina to display their bodies and to declare their need to be cared for, but at the same time enable Tina and other women like her who are over 30 and/or who already have children to access possibilities for marriage and perhaps children at a later age than is the cultural norm. This is not unlike the context for innovations in internet dating outlined in relation to Uganda and, as we will see, is also relevant in China and Turkey. Public norms are not always the only story when examining innovations in intimacy. Once again, thick accounts and explanations of social change are required.

China

China's constitution guarantees equal rights between men and women, and Chinese women have enjoyed gains in their health and literacy rates. Yet, there is still a marked gender gap in income, government positions are regularly posted as 'men preferred' or 'men only', and efforts to protest sexual abuse and harassment have been subject to government suppression. In this context, while gender polarity and inequality are less obviously marked on the Chinese than on the Argentinian dating site we selected, gender distinction remains clearly evident in the Chinese profiles. In China, women are still expected to follow the orders of a husband upon marriage and be obedient to him (Catalyst 2020; Chang 2020; China Power Team 2020; Lewis 2020; Qin 2019) and the profiles reflect such gender differentiated positionings in many ways. As mentioned earlier, China's values in relation to heterosexual gender relations are not bound by religion but remain 'traditional' in other restrictive ways and hence we dispute the location of China on the WVS Cultural Map.

On the selected Chinese internet dating site, *China Love Cupid*, there is unsurprisingly little evidence of religion constraining preferences for potential partners. This is a point of differentiation from even those countries which are largely secular in orientation. Nevertheless, there remain other significant restraints upon innovations in developing intimate relationships. In both our general survey of profiles and the closer attention to selected instances, gender differentiation is a notable feature of self-presentation. Women employed photos representing themselves in a comparatively well-covered 'modest' fashion with very few examples of bodily display or overtly sexy poses. Most often, the Chinese women daters provided close-up headshots and sometimes head-and-shoulder shots in which they tilted their heads to one side and smiled coyly. Very commonly these women were university educated but they presented themselves in a demure, unassuming way.

This mode of presentation is not evident in relation to men's profiles, particularly those of older men, in which men provided photos at more of a distance from the camera, usually of head-and-upper-body shots or full-length shots facing the camera. Men's profile photos almost always showed them in formal business suits, often unsmiling, and rarely outdoors. The common presentation of masculinity on the China site may be contrasted with that which is found in the UK and Australian sites, in that these Chinese male daters often represent themselves in a comparatively serious, formal, professional way almost devoid of outdoor or sporting references. While self-descriptors like 'relaxed' and 'easy-going' are a feature of men's accounts in Britain and Australia, almost the opposite can be found in relation to the selected Chinese dating site.

Though gender differentiation signalling different evocations of authority is an element of the Chinese profiles, one of the most outstanding constraints on unconventional directions arises around the importance of family life and parents, as well as the heteronormative pressure to marry and have children by a certain age, recalling the pressures found in relation to Uganda and Argentina. Dating is predominantly linked to marriage or serious relationship in China and there is considerable parental and social influence brought to bear to constitute dating as a preliminary to permanent familial bonding rather than as an end in itself (Blair and Madigan 2019; Hendricks 2018; Wong 2014). The continuing emphasis in China on obedience to hierarchically organised social structures and relationships, loyalty, familial piety, collectivity and societal harmony encourages support for cultural norms regarding intimate relationships rather than supporting personal individual decision-making. This limits heterodoxical innovations (Evason 2015; Wang and Nehring 2013). The relentless pressure to conform to heteronormative requirements is such that men and more particularly women are deemed to be 'left-overs' if they do not marry before their late twenties (Lewis 2020; Budden 2017; Ng and Ning 2018; Guo 2017). This characterisation is similar to Ugandan conceptions of unmarried childless women as 'useless'.

All the same, the comparative scarcity of women in China offers the possibility of at least some innovative directions for women. In the national context of China, daters must confront a hefty gender imbalance. In 2018, there were around 31 million more men than women in the country (Textor 2020; Budden 2017). This imbalance has led to a rise in kidnapping and trafficking in women as well as buying brides from within China and beyond (Foreign Correspondent 2020). Although we may wish to avoid demographic determinism, the context of such a disparity appears to give Chinese women daters more power in the internet dating world. Women (over 27) who have been deemed 'left-over' have seemingly become more valued. On the sites we examined it is common to find Chinese men's profiles declaring that they are seeking women in age ranges which extend past their own age and indeed women who are often significantly older than themselves (for example, 10–25 years older), and/or announcing as does 45-year-old Lin that he does not care about age but only that there is mutual liking. This is a significant reversal of all the other dating sites we have examined. Male daters around the world are strongly inclined to assert strong preference for younger women and this is confirmed in numerous scholarly examinations (see Chapter 5). Thus, expressing a willingness to view older women as potential partners represents an unusual turnaround in men's culturally established authoritative positioning and opens up new partner opportunities for women.[16] Perhaps it is no surprise in this setting to find that, despite the continuing importance of filial piety such that both women and men express a wish to involve and comply with parents in partner choice-making, yet many do not (Wang and Nahring 2013). Internet dating provides a means to meet norms regarding marriage, children and women's social role, and simultaneously redirect these norms away from parental or other social restrictions. It offers a means to elude normative imperatives in changing social circumstances.

Turkey

In Turkey, this mixture of normative constraints and unconventional practices is also clearly present. Online dating has a transitional status in Turkey which in many ways reflects the nation's location as both a 'traditional' country and a non-Western country which displays aspects of so-called 'modernising' change. This discordant complexity is evident in the middling but still comparatively traditional position of Turkey in the WVS Cultural Map. Such internal dissonance is also manifest in national responses to the internet. On the one hand, over 75 per cent of the population aged 16–74 use the internet, and 60 per cent aged 18–54 are reported to engage in internet dating (Daily Sabah 2019; Kemp 2020b; Uysal 2019). On the other hand, strongly negative attitudes towards online dating—particularly with regard to longer-term relationships and marriage—are common (Hürriyet daily news 2019). The conception of the internet as a potentially immoral space shapes engagement with online dating, especially for women (Miller et al. 2016: 117) In this sense, those who undertake internet dating in Turkey are already engaging in an innovative practice regarding intimacy. This goes some way to explaining some specific features of the dating pool on the Turkish dating site we selected—that is, *LoveHabibi*.

The marked conflicting tensions between traditional and 'modernising' values and practices have a substantial impact upon heterosexual gendered and familial relations. Some mass media and other institutions such as the education system in Turkey popularise equality between men and women, contact between the sexes, and individual decision-making in partner choice. However, cultural norms for the most part involve the expectation that men be dominant over women, that men and women should have largely separate lives and that parents should shape or at least monitor any relationships between them with the goal of marriage. Despite certain legal improvements in women's position, in recent years a decline in secular discourse in political and civil life has generated a more hostile environment for social change around gendered heteronormativity with a shift towards constituting women as wives and mothers and relatedly prioritising the family (Bakan and Tunçel 2019; Mutluer 2019; Bartkowski et al. 2018; Kocamaner 2018; Cindoglu and Unal 2017; Hürriyet daily news 2015).

On a number of online dating sites around the world, racialised/ethnic filters have been discredited and sometimes abandoned or, as in Australia, are largely deemed as of limited significance in terms of partnership (Adewunmi 2010; see also Chapters 4 and 5 and earlier in this chapter). However, on a range of dating sites available in Turkey such as *TurkishPersonals*, *Tinder*, *Siberalem*, *Muslima* and *Happn*—as well as on *LoveHabibi*—there is a comparatively striking emphasis on filters. These filters compel development of profiles around categories of race/ethnicity, but additionally on the basis of ancestry, religion and associated dietary restrictions. Whereas on the Chinese internet dating site the most prevalent account of religious affiliation states 'no religion', on the Turkish sites self-presentation is strictly fixed by cultural specifications and religious orientations are manifestly featured.

On the dating site *LoveHabibi*, there are certain gendered and age-based differences in self-presentation in photos on the profiles. The younger men's profiles are not formal in stance or clothing styles. Older men (40+) on other sites like *Muslima* and *Siberalem* present themselves with more serious expressions, often unsmiling, with formal clothing more likely and very often with moustaches or other modes of facial hair. This is perhaps indicative that *LoveHabibi* has a somewhat younger and less traditional membership—a point to which we will return. However, the Turkish men rarely present themselves in as formal a way as those on the China site. Women's profiles almost invariably show them in close-up and as smiling. Most women, particularly younger women (20s–30s) are dressed in unrevealing everyday casual wear and do not present in a sexual way. However, there is also a clear divide between the prevalence of this casual mode of self-representation and that undertaken by some women who wear headscarves and are in clothing which completely covers neck and shoulders.

We see here an evident sign of the tensions between those values which are associated with more secular and expressive self-presentation and increasingly politicised so-called 'traditional' values associated with women dressing 'modestly' to demonstrate their commitment to Islam and often also as a statement rejecting Western modernism (Carle 2004: 63). Given the constitutional principal of official secularism still in place in Turkey, wearing of the hijab headscarf was banned for many decades in public sector workplaces. The ban was lifted in 2013 (Smith 2013). Hence, this mode of dress is politically and socially significant in Turkey (Sari 2019). Profiles which show women wearing headscarves explicitly and consistently seek marriage. The issue of women's modesty also sometimes appears on men's profiles. For example, Berkay (48) characterises himself as a 'good Muslim gentleman', seeking a 'decent modest Muslim lady', whereas Fatih (27) stresses his 'moderate values' and makes a point of his interest in finding an 'open-minded person' with whom he could share ideas.

Internet dating on *LoveHabibi* appears overall as favouring 'modernising' directions. As is the case in Uganda, those embarking on dating online in Turkey are comparatively educated, with most being at least university graduates, and employed in full-time jobs. In short, internet dating in Turkey is largely undertaken by a relatively affluent elite (Uysal 2019). In the light of the emphasis upon religiously based 'family values' since President Erdoğan's AKP party came to power in 2002—which involve rejection of gender equality, promotion of early marriage and shaming of childless women—culturally valued norms around intimate relationships are increasingly weighty (BBC News 2019; Kocamaner 2018). What is interesting in this context is that members of the *LoveHabibi* dating site, both women and men, very commonly designate themselves as specifically holding 'moderate' or 'liberal' values. Like daters in Uganda, Argentina and China, online users in Turkey are able to meet compelling social demands concerning marriage, children and the place of women as potential wives and mothers, yet at the same time reorient these demands such that social restrictions do not entirely smother more idiosyncratic and diverse possibilities. As Miller et al. state, the availability of internet interaction between men and women in Turkey has had a liberalising effect in creating 'unprecedented possibilities for cross-gender contact and the fulfilment of romantic aspirations' (2016: xiv; see also Al-Heeti 2019; Green 2014).

Exploring innovations across seven countries

When we bring to the fore our research agenda of attending to heterodoxical (non-normative) practices within internet dating there do appear to be signs of innovation. In this chapter, the centre of our attention was dating profiles in each country associated with selected dating sites. Within that field of investigation, we focussed upon one or two notable directions in each country in relation to innovation, enabling emphasis upon different contextualised themes—that is, shifting gender differentiation (UK), expanding marriage parochialism (Uganda), regression from more equitable public norms (Denmark), shifting sexual practices (Australia), expanding gendered age opportunities (Argentina), shifting gender power relations and evading familial restrictions on intimate partnership/marriage (China) and eluding political/religious and familial constraints on cross-gender contact and partner choices (Turkey). It is important to note that these are by no means the only innovations to be found in these countries, as the more extensive accounts of daters' experiences in the UK and Australia in the preceding chapters make clear. Rather, in this chapter our pilot and schematic studies of profiles deliberately concentrate our viewpoint for exploring possible innovations. Indeed, as we have observed before, examination of profiles may narrow that viewpoint to the least amenable material in internet dating in which to locate nonconformity (see also Chapter 3). Moreover, the innovations we have pointed out in each country may of course be found in countries other than the ones with which we have associated them. Our aim here is simply to draw attention to how internet dating can reveal sometimes unexpected possibilities across the globe.

We note, in the context of this analysis of sites and profiles, that there are at least three main tropes in terms of global nodes for change—associated with shifting gender subjectivities and relations, cultural restrictions such as familial requirements, and constraints regarding homogamy (see Yancey 2007). For both heterosexual women and men in this international frame, the availability of less parochial options in partner choice and for actively engaging in considering a range of options, necessarily offers some disruption to what in many countries amounts to a compulsorily homogamous and limited terrain for intimacy. In Uganda and China, as in Australia, internet dating does open up a wider field for connection, desire and love that is perhaps particularly advantageous for those whose space for action has been typically constrained by any, some or all of the following—that is, family, community, religion, culture, nation, age, hetero-sex, hetero-coupledom and gender.

Notes

1 See also the terminology of 'queer geographies' which similarly draws upon critical connections between place and space and the production of embodied subjectivities, practices, social/sexual connections, institutional arrangements, societal power relations and social justice (Mayhew 2015: 409). There is also a similar body of literature on 'emotional geographies' (see Davidson et al. 2007).

2 Our collaborators in the 2014–2016 pilot study were Caroline Wamala Larsson (previously Caroline Wamala), at the Department of Computer and Systems Sciences, and Head of Research, SPIDER Centre, University of Stockholm, Sweden, and Katherine Harrison, Gender Studies (TEMAG), Department of Thematic Studies (TEMA), Linköping University, Sweden. We draw in this chapter on the account of the collaborative pilot study published in Beasley et al. 2018. That 2018 publication was co-authored by all four authors. While most of the work in this chapter is new and developed for this book, the earlier collaborative thinking influenced this chapter's broad outlines and there is some limited utilisation of previous content.

3 The Global Gender Gap Index measures gender-based gaps based on four key dimensions: Economic participation and opportunity, Educational attainment, Health and Survival and Political empowerment (WEF 2020: 5).

4 The UK's median income has been on the rise but most of this has been a result of the rise in average income for the richest fifth. In terms of wealth inequality, the richest 10 per cent of households hold 44 per cent of all wealth in the country (The Equality Trust 2019).

5 See Bridges and Pascoe for one theorisation of such 'hybrid' assemblages of gender involving normative and heterodox features (2014).

6 This understanding of (unmarried) childless women is by no means limited to Uganda, as shall be seen when we turn to Argentina and China. It is made explicit in the Italian saying, 'La Dona senza fioi come "na vegan morta" ' (A woman without children is as useless as a dead vine) (Valmorbida 2018; 41).

7 Local newspaper.

8 Local television dating show.

9 We refer here to Denmark proper—that is, our outline and discussion does not include semi-autonomous Greenland or the Faroe Islands.

10 Denmark is marked by a remarkable population homogeneity (Athanasiadis et al. 2016; Barry and Sorensen 2018; Hedetoft 2006). Denmark proper is the smallest Scandinavian country in continental Western Europe, (Anderson 2020; Winsor 2014) with an area of about 43 thousand square kilometres. This land area may be compared with the UK (243 thousand km2), Uganda (241.5 thousand km2) and Australia (7.7 million km2; Worldpopulationreview Countries by Area 2020). It has a relatively small population. For instance, if we compare our four pilot study countries they rank as follows with regard to population by country: UK Ranking 21/0.87% of world population; Uganda Rank 32/0.59%; Australia Rank 55/0.33%; Denmark Rank 111/0.07 (Worldpopulationreview Population by Country 2020). Nevertheless, despite the limitations of its size in spatial and population terms, Denmark has a high level of economic prosperity and presently has one of the lowest debt-to-GDP ratios in Europe (NordeaTrade 2020; MakingItMagazine 2013).

11 The Equal Measures2030 SDG Gender Index is based upon a range of indicators making use of 14 of the 17 official Sustainable Development Goals which aims to provide a means to register links between the impact of gender relations and poverty.

12 Fembots are humanoid robots resembling a woman in appearance or computer progammes designed to present as women. In internet dating sites, fembots take the form of fake profiles, usually to encourage male users that there are a greater

number of females on the sites than is actually the case and to lure them with photos of desirable women.

13 Along with consistently high endorsement of multiculturalism, the Scanlon Foundation Mapping Social Cohesion national survey found, for example, that 81 per cent disagreed with discrimination on grounds of race or ethnicity (Markus 2019: 3–4).

14 *RSVP* conducted a survey of over 2,400 members online in December 2015 (RSVP Date of the Nation Report 2015). This material resulted in five dating tips offering guidance arising from members' comments posted by the *RSVP* team on January 15, 2016 (RSVP team 2016). However, now the report itself and the subsequent posting are not accessible and appear to have been archived. The report and associated material were initially cited in a paper presented at the University of Edinburgh on 22 September, 2016).

15 The two remaining continents, Antarctica and North America, are either irrelevant (in that people in Antarctica are nationals of other countries) or already have a significant, indeed dominant place in studies of internet dating.

16 The disparity between numbers of men and women and associated problems faced by men seeking potential partners in China may be particularly acute—as is the case in India—for marginalised poor and/or rural men (Chaudhry 2018), though clearly it has also had an impact upon the comparatively advantaged men on the *China Love Cupid* dating site.

Chapter 8

Conclusion

This book aims to investigate the question of whether online dating has altered the dating landscape 'for the better', as Finkel et al. put it (2012: 3). However, we have both more specific and broader concerns than consideration of internet dating *per se*. Rather, our key question and guiding argument in the book revolve around a socio-political agenda. We ask, *'Does heterosexual internet dating as a new technology of intimacy offer any opportunities for enhancing social equality, specifically for moving beyond the constraints of gendered heteronormativity and advancing gender equality?'* Rather than examining whether internet dating produces an improved dating setting, we focus upon the possibility of whether dating online can generate progressive improvement in *heterosexual gender relations*. To determine what change internet dating might bring to these relations, we have outlined what we understand as normative and what we conceive as progressive non-normative possibilities. Our argument throughout the book thus crucially revolves around offering an original contribution to theorising social change, through clarification of what and how social change occurs in intimate relationships.

This principal concern with theorising social change is focussed upon intimacy and internet dating as a technology of intimacy, but is further defined by a deliberate concentration upon heterosexual accounts of internet dating because of a theoretical interest in exploring signs of social change within 'the realm of the dominant' (Beasley et al. 2012: 84). We wish to go beyond the common inclination to seek change at the social margins and to explore possibilities that might challenge notions of the homogeneous and impermeable character of the mainstream. Our concern here is to avoid conceptions of social change in which the marginalised are constituted as necessarily and always the vanguard of refashioning society and the majority are presumed to be excluded from social transformation. To us this seems a remarkably circumscribed, oddly hierarchical and profoundly undemocratic account of the subtleties of social change. Consequently, we explore a critical analysis of gendered heterosexuality in which it is decidedly not viewed as an undifferentiated and unchanging monolith which is inevitably and eternally heteronormative (Beasley et al. 2012 passim).

Instead, in this book we draw attention to the micro-politics of social change to emphasise non-normative directions in gendered heterosexuality. Attention to the micro-political, rather than to the more usual concern with activities at a macro-scale, seems to us more in keeping with our investigation of intimate sociality and enables a more nuanced approach to how change arises and takes shape. Talking about the micro-political aspects of sociality including those deemed 'private' (such as the development of intimate relationships) allows us to register what is often rendered politically invisible in discussions of social change. In short, the book and its foregrounding of the voices of internet daters, in conjunction with other sources, aims to flesh out understandings of social change in ways that value the everyday, small scale and personal. We wish to underline that social change is not just to be found in the vast heave of world-changing events and extraordinary turning points, but also in delicate cumulative small alterations which, though precarious and unpredictable at first, take on force and weight over time. Moreover, the subtle droplets of micro-politics may well be the more likely way that social change arises in the interactions of intimate sociality. Our intention in this focus on the nuanced possibilities of intimate connections is to bring to the fore not only the uneven character of heterosexual gender relations but importantly to employ this field as a means to give life to the inconsistent and volatile character of social change itself.

In terms of the content of the book, we initially established a theoretical and methodological framework for our investigation of internet dating in the Introduction (Chapter 1) and then outlined our main concern with the experiences of internet daters in Britain and Australia. In the second chapter, we established where the boundaries of heterosexual gender norms in internet dating might be found and hence the point at which innovations might begin. We did this by examining the 'nutter' narrative. The third chapter, by contrast, outlined the terrain of new norms emerging within internet dating and examined to what extent they unsettle heterosexual gender hierarchies. Contestations around etiquette indicate the kinds of reflexivity involved in navigating social innovation. In Chapter 4, we explored how this reflexivity is by no means merely cognitive and rational but also embodied and emotional, enabling internet daters to navigate towards more heterodox (non-normative) possibilities in ways that are not merely risk-laden and fearful, but sometimes enjoyable and exciting. Chapter 5 further developed our exploration of heterodoxical directions in online dating by focusing on what is arguably its raison d'être—chemistry. In response to its centrality, we investigated its dangers, delights and opportunities. Having spent the first part of the book examining heterodox possibilities by outlining key aspects of heterosexual internet dating, in the final two chapters (6 and 7) we assessed how these possibilities might differ with age and with geographical location. Chapter 6 examined the experience of older internet daters to find that they provide ways in which to re-conceptualise gender, sexuality and social change. Finally, drawing on a collaborative pilot study (Beasley et al. 2017) and a further schematic study, in the seventh chapter we expanded our main investigation of daters in the UK and Australia by adding attention to a further five countries. This broader investigation was undertaken through a narrowed focus on dating sites and profiles. In attending to geographical difference, race/ethnicity and culture are revealed as critical to understanding social change. Examining contexts including Britain, Uganda, Denmark, Australia, Argentina, China and Turkey allows us to consider the potential of our theoretical model of heterodoxy to be considered within and beyond 'Western' contexts. Last, in this Conclusion we aim to draw together all of these materials to provide a condensed account of the central question, guiding argument and content of the chapters, as well as noting novel features of the book. We end with summarising the surprising array of heterodoxical innovations we have found. After all, surveying and/or discovering, and interrogating these innovations in intimacy is what really matters about this book.

And so to those heterodoxical innovations. There is abundant evidence of the micro-politics of change in intimate sociality in this book—that is, of divergent, transgressive, subversive and dissident (if not of heretical) directions. In *Chapter 2* we mention, for example, that, even in the arena of narratives which are primarily intended to reproduce heteronormativity and in particular gender normativity, women do nevertheless sometimes initiate contact with a date, despite the problems attached to being considered too forward, and employ delaying tactics to take some control over the pacing and progression of intimacy which enables *divergence* from norms of feminine passivity and masculine leadership. In *Chapter 3* we draw out these innovations regarding women instigating dates and sexual activity, as well as undertaking practices intended to retain a degree of control over intimacy, such as by deferring offline meetings. In that chapter, several other primarily *divergent* innovations are outlined, though some verge on *transgression* and a few are *subversive*. These innovations include the strategic—even if somewhat questionable—use of petty dishonesty in which women attempt to respond proactively to being valued on the basis of their appearance, daters collaboratively assisting others to connect with possible partners, group dating at singles events, women becoming less concerned with men's financial position and more inclined to share costs of dating, early moves to have sex in spite of continuing concerns about being over-familiar too soon and about casual sex, and equivocation over monogamy and exclusivity.

In *Chapter 4*, we questioned in greater detail common approaches to internet dating which concentrate upon marketisation and risk assessment. This critical perspective enabled us to turn to a range of innovations including resistance to marketisation and positive responses to 'risk', as well as attending to the novel development of heightened emotional reflexivity. Moreover, we outlined intriguing shifts away from relationship inequality (especially with regard to lessening acceptance of inequitable homogamy), as well as shifts away from monogamy towards mundane polyamory and the growing take-up of casual sex, in particular by women. Such innovations involve substantial challenges to gendered heteronormativity, from *divergent* to *transgressive* and occasionally *subversive* possibilities.

There are further signs of heterodoxical non-conformity in *Chapter 5*, where we considered what might be viewed as somewhat more radical forms of novel practices including varieties of *subversion* but blurring into more challenging directions such as intentional *dissidence*. These refashionings of normativity include subversions of gender normativity such as women taking up more proactive engagements with internet dating and, in spite of gender stereotypic assumptions regarding men's preference for much younger women, men's actual practices taking up other pathways. Dissident shifts involve confounding assumptions regarding hetero-sex and hetero-coupledom by evasion, delay or repurposing of internet dating towards personal development, preparatory rehearsal, and turning to options including non-normative alt-platonic coupledom.

As we turned to older dating in *Chapter 6*, it is evident that the mere undertaking of online dating by older adults in the face of ageism entails heterodoxical innovation. This chapter rejects the usual focus on young people which is also evident in the scholarly literature. While these older adults share many similar unconventional directions with younger people, they also undertake *subversion* of gender norms. Older women, for example, show signs of non-normative practices notably in so-called Cougar dating. Older men adopt more mutual and communicative models of hetero-sex and hetero-coupledom which involve considerable shifts with regard to masculinity. Both men and women undertake *dissident* directions by diverting internet dating into options like cross-sex friendship. However, perhaps one of the most intriguing and unconventional possibilities associated with older internet daters lies in the development of novel sexual repertoires. In this unshackling from heteronormative constraints, older internet daters may be unexpectedly viewed as at the forefront of confronting gendered heteronormativity.

Chapter 7 focussed upon dating sites and profiles across seven countries, Nevertheless, it has a narrowed perspective in terms of attending to social change. Because the focus is upon public self-presentation in internet profiles rather than the diversity that might arise in myriad private practices, it may be an aspect of internet dating least amenable to innovations. However, even here, whatever the national cultural context in each country there are notable instances of heterodoxical possibilities. These range from signs of rejection of aggressive masculinity and a seeking for less 'traditional' or patriarchal male partners, to divergence from public discourses around equality and monogamy. Social change takes many forms, is highly variable in its scope and effects and is not always to be found where it might be expected.

This is certainly the case in relation to social change when the focus is upon intimacy in the setting of internet dating. There is considerable evidence of reiteration of unpleasant and constraining aspects of heterosexual gender relations. For example, as Londoner Anna finished playing pool with her new internet date, he turned to her and said, 'If I was to kill you, I'd probably cook you as I'm a chef.' She listened in horror as her date described how he would do this. Anna notes, '[a]fter that, I stopped going on online dates for a long time' *(Anna, UK*—Evans 2020*)*. Unfortunately, Anna is not alone.

Yet, Carol has had a very different experience.

> I don't think I can call myself an expert, but for the last few years I've been 'dating online' pretty successfully: just having fun, meeting cool and interesting people, having craziest conversations in my life, getting short and long term boyfriends, falling in love, falling apart, even meeting a soulmate . . . Of course, not all of these experiences have been joy and fun, but I can't complain.
>
> *(Carol, UK, dating site blog, no date)*

And, notwithstanding Bruce's view of internet dating as largely depressing, when all is said and done he is jubilant.

> Did online dating for a few years as I live in a small country town and there is no dating scene here to speak of. Met a few women . . . but usually didn't get any further than 1 or 2 dates at the most. Got the 'you're a nice guy, but I don't see us being anything more than friends' line more than I would have liked to have heard. It got quite depressing for me at times, especially later on when I had to wonder what I was doing wrong or maybe I was just meant to be on my own.
>
> After another fizzled attempt . . . I decided to get back on the horse as the saying goes . . . I met a Chinese woman . . . I hadn't considered dating someone that was Asian before but she was cute and seemed nice . . . so I thought 'why not?' . . . Fast forward to now and we're about to celebrate our 2 year anniversary and are both very happy . . .
>
> So what is my opinion of internet dating? Most of the time, it sucked. Hard. But it only takes meeting that one special person to make all of the bad times I went through totally worth it in the end.
>
> *(Bruce, Australia, online dating story compilation, 2014)*

As can be seen from these examples, internet dating can be a horror story, an unforeseen opportunity or even a saviour. Whatever the case, in each instance it is important to be alert to possible social innovations which refashion gendered heteronormativity and advance gender equality. Intimacy does not necessarily follow the cultural rules. It may be at odds with them (such as in the case of the dating site, *Victoria Milan*, in Denmark) or may divert or redirect them (for instance, in the examples of avoidance of parental involvement or redirected monogamy). The point is that possibilities for social change, in relation to intimacy and other domains of sociality, arise unpredictably and are not always noticed. For instance, a commentator drawing on the 2010 *Date of the*

Nation Report by Australian dating site, *RSVP* (undertaken by an independent survey company), is surprised to find that Australian internet daters are strikingly traditional in seeking love and marriage.

> We might be using new technology to find that special person, but the . . . survey found Aussies still hold traditional values when it comes to love and marriage.
>
> Three quarters of Australians want to spend their lives with one person and more than half see marriage as an important institution.
>
> [The survey] found more men (78 per cent) than women (70 per cent) aspire to settling down and spending their life with one partner and 54 four per cent of single men and 46 per cent of single women have aspirations to marry.
>
> *(Anonymous 2010*, see also Feeney 2010*)*

However, we notice another story here—that is, contrary to usual expectations, men engaged in internet dating are apparently more invested than their women peers in cultural norms regarding monogamous, permanent and formal commitment to an intimate partner. Internet dating, intimacy and social change do not necessarily follow tried and true paths. That is what we believe is valuable about this book. Indeed, we hope that it offers grounds for optimism to all those of us who seek social equality.

References

Note: All web addresses were accessed between 2014 and 2021. Only two are presently unavailable.

ABC News (2017) 'Fact check: Does the data show that income inequality has fallen?' *ABC News Fact Check*, Tuesday 29 August [https://www.abc.net.au/news/2017-08-16/fact-check-income-inequality-has-it-improved/8804976].

Abgarian, A. (2018) '14 men tell us why they want more women to initiate a date—and how.' *Metro*, 13 May [https://metro.co.uk/2018/05/13/14-men-tell-us-why-they-want-more-women-to-initiate-a-date-and-how-7521480/].

Abramova, O., Baumann, A., Krasnova, H. and Buxman, P. (2016) 'Gender differences in online dating: What do we know so far? A systematic literature review.' 9th Hawaii International Conference on System Sciences (HICSS), Koloa, HI: 3858–3867.

ABS (Australian Bureau of Statistics) (2013) 'Perspectives on Migrants', March [https://www.abs.gov.au/statistics/people/people-and-communities/perspectives-migrants/latest-release].

Acemoglu, D., Johnson, S., Robinson, J. A. and Yared, P. (2009) 'Reevaluating the modernization hypothesis.' *Journal of Monetary Economics* 56: 1043–1058.

Adams, M. S., Oye, J. and Parker, T. S. (2003) 'Sexuality of older adults and the internet: From sex education to cybersex.' *Sexual and Relationship Therapy* 18(3): 405–415.

Adewunmi, B. (2010) 'Racism on dating sites.' *The Observer*, October 29 [https://www.observer.ug/index.php?option=com_content&view=article&id=10701].

Adler, R. P. (1996) 'Older adults and computers: Report of a national survey.' SeniorNet: San Francisco, CA [http://www.seniornet.org/php/ default.php?Page ID=5476].

AgeUK (2016) 'The internet and older people in the UK—Key Statistics', July [https://www.ageuk.org.uk/globalassets/age-uk/documents/reports-and-publications/reports-and-briefings/active-communities/rb_july16_older_people_and_internet_use_stats.pdf].

Ahmed, S. (2006) *Queer Phenomenology: Orientations, Objects, Others*. Durham and London: Duke University Press.

Ahmed, S. (2004) *The Cultural Politics of Emotion*. London: Routledge.

Aiken, J. (2016) 'Is it ok to date more than one person?' datehub by RSVP (site advice) [http://www.datehub.com.au/single-post/2016/05/09/Dating-more-than-one-person].

Alarie, M. (2020) 'Sleeping with younger men: Women's accounts of sexual interplay in age-hypogamous intimate relationships.' *The Journal of Sex Research* 57(3): 322–334.

Alarie, M. (2019) '"They're the ones chasing the Cougar": Relationship formation in the context of age-hypogamous intimate relationships.' *Gender & Society* 33(3): 463–485.

Alarie, M. (2017) *Beyond the "Cougar" Stereotype: Middle-aged Women's Experiences of Intimate Relationships with Younger Men.* PhD dissertation, Sociology: Magill University.

Albo, B. (2017) 'Top dating blogs you're not reading—but should be', February 15 [https://www.liveabout.com/top-dating-blogs-youre-not-reading-but-should-be-1022103].

Albury, K. McCosker, A., Pym, T. and Byron, P. (2020) 'Dating apps as public health "problems": Cautionary tales and vernacular pedagogies in news media.' *Health Sociology Review*, DOI:10.1080/14461242.2020.1777885.

Alexander, L. (2018) 'What's it like to enter the online dating scene in middle age?' *The Sydney Morning Herald*, June 16 [https://www.smh.com.au/lifestyle/life-and-relationships/what-s-it-like-to-enter-the-online-dating-scene-in-middle-age-20180612-p4zl02.html].

Al-Heeti, A. (2019) 'Beyond Tinder: How Muslim millennials are looking for love.' *C/Net*, April 10 [https://www.cnet.com/news/tinder-minder-muzmatch-eshq-how-muslim-millennials-are-looking-for-love/].

Allain, K. and Marshall, B. (2017) '"Buff boys" with brooms: Shifting representations of masculinity in Canadian men's curling.' *International Journal for Masculinity Studies* 13(3): 1–17.

Allen, L (2003) 'Girls want sex, boys want love: Resisting dominant discourses of (Hetero) sexuality.' *Sexualities* 6(2): 215–236.

Alterovitz, S. S. R. and Mendelsohn, G. A. (2013) 'Relationship goals of middle-aged, young-old, and old-old internet daters: An analysis of online personal ads.' *Journal of Aging Studies* 27(2): 159–165.

Alterovitz, S. S. R. and Mendelsohn, G. A. (2009) 'Partner preferences across the life span: Online dating by older adults.' *Psychology and Aging* 24(2): 513–517.

Among, I. and Munavu, M. M. (2019) '"We want to be heard": The voices of Uganda's young people on youth unemployment.' World Bank Blogs, April 30 [https://blogs.worldbank.org/nasikiliza/we-want-to-be-heard-the-voices-of-ugandas-young-people-on-youth-unemployment].

Andersen, J. G. and Shamshiri-Petersen, D. (2016) 'Attitudes towards gender equality in Denmark, Sweden and Norway.' Paper presented at the 2016 Annual meeting of the Danish Political Science Association, October 27–28, Vejle, Denmark [https://dpsa.dk/papers/ShamshiriPetersen%20Goul%20Andersen.Gender%20equality%20attitudes.DPSA%20paper(1).pdf].

Anderson, E. (2012) *The Monogamy Gap: Men, Love, and the Reality of Cheating.* New York: Oxford University Press.

Anderson, M., Vogels, E. and Turner, E. (2020) 'The virtues and downsides of online dating.' Pew Research Center, February 6 [https://www.pewresearch.org/internet/2020/02/06/the-virtues-and-downsides-of-online-dating/].

Anderson, R. (2016) 'The ugly truth about online dating.' *Psychology Today*, September 6 [https://www.psychologytoday.com/au/blog/the-mating-game/201609/the-ugly-truth-about-online-dating].

Anderson, S. V. (2020) 'Denmark.' *Britannica*, Countries of the World, September 1 [https://www.britannica.com/place/Denmark].

AnKee, A. and Yazdanifard, R. (2015) 'The review of the ugly truth and negative aspects of online dating.' *Global Journal of Management and Business Research* 15(4) [https://www.journalofbusiness.org/index.php/GJMBR/article/view/1700].

Anonymous/PC Tech (2020) 'Uganda: Internet penetration stands at 37.9%, with 23m users.' PC Tech, February 3 [https://pctechmag.com/2020/02/uganda-internet-penetration-stands-at-37-9-with-23m-users/].

Anonymous (2010) 'Online love leads to marriage.' *The Sydney Morning Herald*, June 30 [https://www.smh.com.au/lifestyle/online-love-leads-to-marriage-20100630-zkpr.html].

Archer, M. S. (2007) *Making Our Way Through the World: Human Reflexivity and Social Mobility*. Cambridge: Cambridge University Press.

Archer, M. S. (2003) *Structure, Agency and the Internal Conversation*. Cambridge: Cambridge University Press.

Aretz, W. Demuth, I., Schmidt, K. and Vierlein, J. (2010) 'Partner search in the digital age. Psychological characteristics of Online-Dating-Service-Users and its contribution to the explanation of different patterns of utilization.' *Journal of Business and Media Psychology* 1: 8–16.

Arioso, L. (2010) 'Personal documents on the internet: What's new and what's old.' *Journal of Comparative Research in Anthropology and Sociology* 1(2): 23–38.

Arndt, B. (2016) 'And so to bed . . . a new sexual etiquette for online daters.' *The Australian*, June 25 [https://www.theaustralian.com.au/nation/inquirer/and-so-to-bed—a-new-sexual-etiquette-for-online-daters/news-story/bf967213fa146fcdd1fec830214450e2].

Arndt, B. (2015a) 'Not looking for "The One".' Dating Advice, *RSVP* website, Australia, 30 July.

Arndt, B. (2015b) 'Getting on and still getting it on: Singles, dates and sex.' *The Sun-Herald, Extra Sunday*, June 21: 228.

Arndt, B. (2012) 'Why women lose the dating game.' *Sydney Morning Herald*, April 22 [https://www.smh.com.au/lifestyle/why-women-lose-the-dating-game-20120421-1xdn0.html].

Asendorpf, J., Penke, L. and Back, M. (2011) 'From dating to mating and relating: Predictors of initial and long-term outcomes of speed-dating in a community sample.' *European Journal of Personality* 25: 16–30.

Athanasiadis, G., Cheng, J., Bjarni, J. V., Jørgensen, F. G., Als, T. D., Le Hellard, S., Espeseth, T., Sullivan, P. F., Hultman, C. M., Kjærgaard, P. C., Schierup, M. H. and Mailund, T. (2016) 'Nationwide genomic study in Denmark reveals remarkable population homogeneity.' *Genetics* 204(2): 711–722.

Australian Bureau of Statistics (ABS) (2018) 'Household use of information technology, Australia, 2016–17', 8146.0, 28 March [https://www.abs.gov.au/ausstats/abs@.nsf/mf/8146.0].

Australian Bureau of Statistics (ABS) (2013) 'Age differences between partners', 4102.0. *Australian Social Trends*, July 2013. [https://www.abs.gov.au/AUSSTATS/abs@.nsf/Lookup/4102.0Main+Features10July+2013].

Australian Bureau of Statistics (ABS) (2009) 'Partners' characteristics', 4102.0. *Australian Social Trends*, March [https://www.abs.gov.au/AUSSTATS/abs@.nsf/Lookup/4102.0Main+Features20March%202009].

Australian Communications and Media Authority (2018) Communications report 2013–14 series. Report 1—'Australians' digital lives'. March. [https://www.acma.gov.au/sites/default/files/2019-08/research-digital-lives-Mar-2015.pdf].

Australian Human Rights Commission (2017) 'Information concerning Australia's compliance with the International Covenant on Civil and Political Rights' [https://humanrights.gov.au/our-work/legal/submission/information-concerning-australias-compliance-international-covenant-civil].

Australian Human Rights Commission (2013) 'Rights and freedoms: Right by right' [https://humanrights.gov.au/our-work/rights-and-freedoms/rights-and-freedoms-right-right].

Australian Human Rights Commission Section 7 (2011) 'Section 7—Protection from discrimination on the basis of sexual orientation—addressing sexual orientation and sex and/or gender identity' [https://humanrights.gov.au/our-work/section-7-protection-discrimination-basis-sexual-orientation-addressing-sexual-orientation#Heading494].

Australian Productivity Commission (2013) 'An ageing Australia: Preparing for the future', Productivity Commission Research Paper, November. Canberra: Commonwealth of Australia.

Australian Productivity Commission (2005) 'Economic Implications of an Ageing Australia', Productivity Commission Research Report, 24 March. Canberra: Commonwealth of Australia.

Bakan, R. and Tunçel, Ö. (2019) 'Turkey's transition from democracy to authoritarianism challenges the feminist movement.' *New Age Islam*, 6 July [https://www.newageislam.com/islam-women-and-feminism/ronay-bakan-and-%C3%B6zlem-tun%C3%A7el/turkeys-transition-from-democracy-to-authoritarianism-challenges-the-feminist-movement/d/119094].

Baker, A. J. and Whitty, M. T. (2008) 'Researching romance and sexuality online: Issues for new and current researchers', pp 34–49. In S. Holland (ed.), *Remote Relationships in a Small World*. New York: Peter Lang.

Bali, M. (2021) 'Cougar characteristics', January 4 [https://metrobali.com/cougar-characteristics-8/#:~:text=Alarie%20proposes%20that%20negative%20reactions, value%20stability%20that%20is%20financial].

Balkmar, D. (2012) *On Men and Cars: An ethnographic study of gendered, risky and dangerous relations*. Linköping Studies in Arts and Science No. 558. Linköping, Sweden: Gender Studies (TEMA), Department of Thematic Studies (TEMAG), Linköping University.

Barnat, N., MacFeely, S. and Peltola, A. (2019) 'Comparing global gender inequality indices: How well do they measure the economic dimension?' *Journal of Sustainability Research* 1: e190016 [https://sustainability.hapres.com/htmls/JSR_1114_Detail.html].

Barnwell, A. (2018) 'Durkheim as affect theorist.' *Journal of Classical Sociology* 18(1): 21–35.

Barraket, J. and Henry-Waring, M. S. (2008) 'Getting it on(line): Sociological perspectives on e-dating.' *Journal of Sociology* 44(2): 149–165.

Barrett, C. (2011) 'Auditing organisational capacity to promote the sexual health of older people.' *Electronic Journal of Applied Psychology* 7(1): 31–36.

Barrett, C., Harrison, J. and Kent, J. (2008) 'Permission to speak. Towards the development of gay, lesbian, bisexual, transgender and intersex friendly aged care services.' [www.matrixguildvic.org.au/publications.htm].

Barry, E. and Sorensen, M. S. (2018) 'In Denmark, harsh new laws for immigrant "ghettos".' *The New York Times*, July 1 [https://www.nytimes.com/2018/07/01/world/europe/denmark-immigrant-ghettos.html].

Barry, N. (2017) 'In Australia, land of the "fair go", not everyone gets an equal slice of the pie.' *The Conversation*, January 27 [https://theconversation.com/in-australia-land-of-the-fair-go-not-everyone-gets-an-equal-slice-of-the-pie-70480].

Bartkowski, J., Acevedo, G., Karakeci, G. and Campbell, F. (2018) 'Islam and support for gender inequality among women in Turkey.' *American Journal of Islamic Social Sciences* 35: 25–55.

Bartlett, A., Clarke, K. and Cover, R. (2019) 'Introduction: Flirting, scandal, intimacy', pp. 1–21. In *Flirting in the Era of #MeToo*. Cham: Palgrave Pivot.

Bateson, D. J., Weisberg, E., McCaffery, K. J. and Luscombe, G. M. (2012) 'When online becomes offline: Attitudes to safer sex practices in older and younger women using an Australian internet dating service.' *Sexual Health* 9(2): 152–159.

Bauman, Z. (2003) *Liquid Love: On the Frailty of Human Bonds*. Cambridge: Polity Press.

Baumeister, R. F. and Leary, M. R. (1995) 'The need to belong: Desire for interpersonal attachments as a fundamental human motivation.' *Psychological Bulletin* 117(3): 497–529.

Baym, N. (2015) *Personal Connections in the Digital Age*, 2nd edition. Malden, MA: Polity Press.

Baym, N. (2010) *Personal Connections in the Digital Age*. Cambridge: Polity Press.

Bazelon, E. (2015) 'The return of the sex wars.' *The New York Times Magazine*, September 10 [https://www.nytimes.com/2015/09/13/magazine/the-return-of-the-sex-wars.html].

BBC News (2019) 'Recep Tayyip Erdogan: Turkey's pugnacious president', 24 June [https://www.bbc.com/news/world-europe-13746679].

Bean, M. (2011) 'Online to in bed: A narrative-discursive exploration of gender, sexuality and heteronormativity in online dating.' Unpublished PhD thesis, University of Auckland.

Beasley, C. (2017a) 'Beyond care and vocabularies of altruism: Considering sexuality and older people', pp. 233–249. In R. Harding, E. Peel and C. Beasley (eds), *Revaluing Care in Theory, Law and Policy: Cycles and Connections*. London and New York: Routledge.

Beasley, C. (2017b) 'What matters in social change? The uncertain significance of caring', The Fay Gale Annual Lecture, for the Academy of Social Sciences of Australia, hosted by University of Sydney, June. [https://socialsciences.org.au/events/fay-gale-lecture-2017-what-matters-in-social-change-the-uncertain-significance-of-caring/].

Beasley, C. (2015) 'Libidinal heterodoxy: Heterosexuality, hetero-masculinity and "transgression".' *Men and Masculinities* 18(2): 140–158.

Beasley, C. (2013) 'The challenge of pleasure: Let's talk about sex in feminist and masculinity studies theorising', pp. 115–126. In M. Fineman and M. Thomson (eds), *Exploring Masculinities: Feminist Legal Theory Reflections*. Farnham, Surrey: Ashgate Press.

Beasley, C. (2012) 'Problematising contemporary men/masculinities theorising: The contribution of Raewyn Connell and conceptual-terminological tensions today.' *British Journal of Sociology* 63(4): 747–765.

Beasley, C. (2011) 'Libidinous politics: Heterosex, 'transgression' and social change.' *Australian Feminist Studies* 26(67): 25–40.

Beasley, C. (2008a) 'The challenge of pleasure: Re-imagining sexuality and sexual health.' *Health Sociology Review* 17(2): 151–163.

Beasley, C. (2008b) 'Re-thinking hegemonic masculinity in a globalising world.' *Men & Masculinities* 11(1): 86–103.

Beasley, C. and Bacchi, C. (2012) 'Making politics fleshly: The ethic of social flesh', pp. 99–120. In A. Bletsas and C. Beasley (eds), *Engaging with Carol Bacchi: Strategic Interventions and Exchanges*. Adelaide: University of Adelaide Press.

Beasley, C. and Bacchi, C. (2007) 'Envisaging a new politics for an ethical future: Beyond trust, care and generosity towards an ethic of social flesh.' *Feminist Theory* 8(3): 279–298.

Beasley, C. and Bacchi, C. (2000) 'Citizen bodies: Embodying citizens—a feminist analysis.' *International Feminist Journal of Politics* 2(3): 337–358.

Beasley. C. and Brook, H. (2019) *The Cultural Politics of Contemporary Hollywood Film: Power, Culture and Society*. Manchester, UK: Manchester University Press.

Beasley, C. and Holmes, M. (2016) 'Internet dating, sexual intimacy and older people', pp. 205–222. In R. Harding and E. Peel (eds), *Ageing and Sexualities: Interdisciplinary Perspectives*. Farnham, Surrey, UK: Ashgate.

Beasley, C., Brook, H. and Holmes, M. (2015) 'Heterodoxy: Challenging orthodoxies about heterosexuality.' *Sexualities* 18(5–6): 681–697.

Beasley, C., Brook, H. and Holmes, M. (2012) *Heterosexuality in Theory and Practice*. London: Routledge.

Beasley, C., Harrison, K., Holmes, M. and Wamala Larsson, C. (2018) 'Innovations in intimacy—Internet dating in an international frame', pp. 89–102. In R. Andreassen, M. N. Petersen, K. Harrison and T. Raun (eds), *Mediated Intimacies: Connectivities, Relationalities and Proximities*. London: Routledge.

Beauvoisin, C. (2015) 'The drama of dating in Uganda.' *The Telegraph*, 22 July [https://www.telegraph.co.uk/expat/expatlife/11753018/The-drama-of-dating-in-Uganda.html].

Beck, U. (1992) *Risk Society*. London: Sage.

Beck, L., Elston, J., Martin, B., Davison, B., Castan, M., Thomson, A., Gardner, M., Bright, D. and Ahmed, A. (2019) 'Australian identity: What does it mean to you?' *A Different Lens*, Episode 11, Lens, Monash University, 22 January [https://lens.monash.edu/@politics-society/2019/01/22/1369645/australian-identity-debate].

Beer, D. and Burrows, R. (2007) 'Sociology and, of and in web 2.0: Some initial considerations.' *Sociological Research Online* 12(5) [http://www.socresonline.org.uk/112/5/17.html].

Belton, P. (2018) 'Love and dating after the Tinder revolution.' *BBC News*, 13 February [https://www.bbc.com/news/business-42988025].

Bennett, K. M. (2007) ' "No sissy stuff": Towards a theory of masculinity and emotional expression in older widowed men.' *Journal of Aging Studies* 21: 347–356.

Beres, M. (2014) 'Points of convergence: Introducing sexual scripting theory to discourse approaches to the study of sexuality.' *Sexuality & Culture* 17(1): 76–88.

Berg, R. C., Molin, S-B. and Nanavati, J. (2020) 'Women who trade sexual services from men: A systematic mapping review.' *The Journal of Sex Research* 57(1): 104–118.

Berlant, L. (2011) 'A properly political concept of love: Three approaches in ten pages.' *Cultural Anthropology* 26(4): 683–691.

Berlant, L. (2000) 'Intimacy: A special issue', pp. 1–8. In L. Berlant (ed.), *Intimacy*. Chicago: University of Chicago Press.

Bernstein, H. (2007) 'Modernization theory and the sociological study of development.' *The Journal of Development Studies* 7(2): 141–160.

Bialik, K. (2017) 'Key facts about race and marriage, 50 years after Loving v. Virginia', June 12. Pew Research Centre [https://www.pewresearch.org/fact-tank/2017/06/12/key-facts-about-race-and-marriage-50-years-after-loving-v-virginia/].

Bingham, J. (2014) 'Love across the divide: Interracial relationships growing in Britain.' *The Telegraph*, 3 July [https://www.telegraph.co.uk/news/politics/10943807/Love-across-the-divide-interracial-relationships-growing-in-Britain.html].

Bishop, L. (2007) 'A reflexive account of reusing qualitative data: Beyond primary/secondary dualism.' *Sociological Research Online* 12(3): 43–56.

Bivens, R. (2017) 'The gender binary will not be deprogrammed: Ten years of coding gender on Facebook.' *New Media & Society* 19(6): 880–898.

Blackwell, D. and Lichter, D. (2004) 'Homogamy among dating, cohabiting and married couples.' *The Sociological Quarterly* 45(4): 719–737.

Blair, S. L. and Madigan, T. J. (2019) 'Dating, marriage, and parental approval: An example of young adults in China.' *Social Science Quarterly* 100(6): 2351–2368.

Bogolyubova, P., Panicheva, R., Tikhonov, V., Ivanov, V. and Ledovaya, Y. (2018) 'Dark personalities on Facebook: Harmful online behaviours and language.' *Computers in Human Behaviour* 78:151–159.

Bolger, D. (2016) 'Race politics: Australian government responses to asylum seekers and refugees from white Australia to Tampa.' Unpublished PhD thesis, School of Humanities and Communication Arts. Sydney, Australia: Western Sydney University.

Bowling Green State University (2012) 'Finding love has no expiration date: People over 60 are fastest growing demographic in online dating.' *Science Daily*, February 11 [https://www.sciencedaily.com/releases/2012/02/120211095051.htm].

Bracewell, L. N. (2016) 'Beyond Barnard: Liberalism, antipornography, feminism, and the sex wars.' *Signs* 42(1): 23–48.

Bradshaw, S. and Howard, P. (2017) 'Troops, trolls and troublemakers: A global inventory of social media manipulation.' Computational Propaganda Project Working Paper. Oxford: Oxford University Press.

Brewer, G. and Hendrie, C. (2011) 'Evidence to suggest that copulatory vocalizations in women are not a reflexive consequence of orgasm.' *Archives of Sexual Behaviour* 40(3): 559–564.

Brickell, C. (2012) 'Sexuality, power and the sociology of the internet.' *Current Sociology* 60(1): 28–44.

Bridges, T. and Pascoe, C. J. (2014) 'Hybrid masculinities: New directions in the sociology of men and masculinities.' *Sociology Compass* 8(3): 246–258.

Brkljačić, T., Sučić, I. and Brdovčak, B. (2017) 'Serious leisure activities and well-being of senior citizens: Case of contract bridge.' *Rocznik Andragogiczny* 24(1): 157.

Brown, M. (2017) 'Is Tinder a match for millennials?' *LendEDU*, 22 March [https://lendedu.com/blog/tinder-match-millennials/].

Brown, M. T. (2009) 'LGBT aging and rhetorical silence'. *Sexuality Research and Social Policy Journal of NSRC* 6(2): 65–78.

Brown, S. L. and Shinohara, S. (2013) 'Dating relationships in older adulthood: A national portrait.' *Journal of Marriage and Family* 75(5): 1194–1202.

Bruch, E. E. and Newman, M. E. J. (2018) 'Aspirational pursuit of mates in online dating markets.' *Science Advances* 4(8): DOI: 10.1126/sciadv.aap9815.

Bruggemann, A. and Pence, S. (2019) 'Empowering women in Buenos Aires and beyond.' *Democracy Speaks*, January 31 [https://www.democracyspeaks.org/blog/empowering-women-buenos-aires-and-beyond].

Bruner, J. (1987) 'Life as narrative.' *Social Research* 54(1): 11–32.

Brynin, M., Longhi, S. and Martinez Pérez, Á. (2008) 'The social significance of homogamy.' Institute for Social & Economic Research 32, October [https://www.iser.essex.ac.uk/research/publications/working-papers/iser/2008-32].

Bryson, L. (1987) 'Sport and the maintenance of masculine hegemony.' *Women's Studies International Forum* 10(4): 349–360.

Budden, R. (2017) 'The gender gap is a big problem in the Middle Kingdom—and its "leftover men" are going to great lengths to find a wife.' *BBC*, Worklife, 15 February [https://www.bbc.com/worklife/article/20170213-why-millions-of-chinese-men-are-staying-single].

Budgeon, S. (2008) 'Couple culture and the production of singleness.' *Sexualities* 11(3): 301–325.

Burkitt, I. (2012) 'Emotional reflexivity: Feeling, emotion and imagination in reflexive dialogues.' *Sociology* 46(3): 458–472.

Burkitt, I. (1998) 'Sexuality and gender identity: From a discursive to a relational analysis.' *The Sociological Review* 46(3): 483–504.

Buss, D. M., and Schmitt, D. P. (1993) 'Sexual strategies theory: An evolutionary perspective on human mating.' *Psychological Review* 100: 204–232.

Butler, J. (2004) *Undoing Gender*. London: Routledge.

Caballero, C. (2019) 'Interraciality in early twentieth century Britain: Challenging traditional conceptualisations through accounts of "ordinariness".' *Genealogy* 3(2) [https://www.mdpi.com/2313-5778/3/2/21/htm].

Cacioppo, J., Hawkley, L., Crawford, E., Ernst, J., Burleson, M., Kowalewski, R., Malarky, W., Van Cauter, E. and Berntson, G. (2002) 'Loneliness and health: Potential mechanisms.' *Psychosomatic Medicine* 64: 407–417.

Calasanti, T. (2009) 'Aging bodies, constructions of masculinities, and the anti-aging industry'. In J. Hearn (ed.), *GEXcel Work in Progress Report, Volume V. Proceedings from GEXcel Theme 2: Deconstructing the hegemony of men and masculinities*. Linköping, Sweden: Linköping University: 27–40 [http://www.genderexcel.org/webfm_send/51].

Calasanti, T. (2004) 'Feminist gerontology and old men.' *Journal of Gerontology* 59B(6): 305–314.

Camacho, M. E. and Reyes-Ortiz, C. A. (2005) 'Sexual dysfunction in the elderly.' *International Journal of Impotence* 17: S52–S56.

Campbell, C. (1987) *The Romantic Ethic and the Spirit of Modern Consumerism*. Oxford: Blackwell.

Cardona, F. (2019) 'The rise of online dating, and the company that dominates the market.' *Visual Capitalist*, March 23 [https://www.visualcapitalist.com/online-dating-big-business/].

Carey, M., Walther, S. and Russell, S. (2009) 'The absent but implicit: A map to support therapeutic enquiry.' *Family Process* 48(3): 319–331.

Carle, R. 2004. 'Hijab and the limits of French secular republicanism.' *Society Abroad* September/October: 63–68.

Carnes, M. (2017) 'Machismo and gender equality in Argentina.' JYAN, blog, Berkley Center for Religion, Peace, and World Affairs, September 28 [https://berkleycenter.georgetown.edu/posts/machismo-and-gender-equality-in-argentina].

Castells, M. (1997) *The Power of Identity. The Information Age: Economy, Society and Culture Vol. II*. Cambridge, MA and Oxford: Blackwell.

Castleman, M. (2012) 'Erection changes after 50: The facts.' *Psychology Today, Australia*, May 1 [https://www.psychologytoday.com/au/blog/all-about-sex/201205/erection-changes-after-50-the-facts].

Castro, N. (2016) 'Argentina, with its "machismo culture", fails in its protection of women and girls.' *Equal Times*, 25 November [https://www.equaltimes.org/argentina-with-its-machismo?lang=en].

Catalyst (2020) 'Quick take: Women in the workforce—China', April 14 [https://www.catalyst.org/research/women-in-the-workforce-china/].

Causa, O., Hermansen, M., Ruiz, N., Klein, C. and Smidova, Z. (2016) 'Inequality in Denmark through the looking glass.' OECD Working Paper No. 1341, 17 November [https://www.oecd.org/denmark/Inequality-in-Denmark-through-the-looking-glass.pdf].

Cavanagh, S. (1997) 'Content analysis: Concepts, methods and applications.' *Nurse Researcher* 4(3): 5–16.

Chakravorti, B. (2017) 'There's a gender gap in internet usage. Closing it would open up opportunities for everyone.' *Harvard Business Review*, December 12 [https://hbr.org/2017/12/theres-a-gender-gap-in-internet-usage-closing-it-would-open-up-opportunities-for-everyone#:~:text=the%20industry's%20products.-,The%20International%20Telecommunications%20Union%20reports%20that%20the%20proportion%20of%20women,she%20might%20face%20additional%20hostility].

Chambers, D. (2013) *Social Media and Personal Relationships*. UK: Palgrave Macmillan.

Chamourian, E. (2017) *Identity Performance and Self Presentation through Dating App Profiles: How individuals curate profiles and participate on Bumble*. Master of Arts in Global Communication. Paris, France: The American University of Paris.

Chang, S. (2020) *Chinese Women, Marriage and Gender: Exploring the idea of women and marriage over time in the context of China*. Major Papers, 109. Ontario, Canada: University of Windsor [https://scholar.uwindsor.ca/cgi/viewcontent.cgi?article=1132&context=major-papers].

Chang, E. S., Kannoth, S., Levy, S., Wang, S. Y., Lee, J. E. and Levy, B. R. (2020) 'Global reach of ageism on older persons' health: A systematic review.' *PloS one* 15(1): e0220857 [https://doi.org/10.1371/journal.pone.0220857].

Chatterjee, C. (2001) 'Can men and women be Friends?' *Psychology Today*, September 1 [https://www.psychologytoday.com/us/articles/200109/can-men-and-women-be-friends].

Chaudhry S. (2018) ' "Now it is difficult to get married": Contextualising cross-regional marriage and bachelorhood in a north Indian village.' In S. Srinivasan and S. Li (eds), *Scarce Women and Surplus Men in China and India. Demographic Transformation and Socio-Economic Development* vol 8. Cham: Springer. [https://doi.org/10.1007/978-3-319-63275-9_5].

Cherkasskaya, E. and Rosario, M. (2019) 'The relational and bodily experiences theory of sexual desire in women.' *Archives of Sexual Behaviour* 48: 1659–1681.

China Power Team (2020) 'Do women in China face greater inequality than women elsewhere?' *China Power*, June 25 [https://chinapower.csis.org/china-gender-inequality/].

Chipperfield, M. (2019) 'Sledging and shorts at the opera—the truth about Australia's classless society.' *The Telegraph, UK*, 18 February [https://www.telegraph.co.uk/travel/lists/australia-classless-egalitarian-society/].

Ciabattari, T. (2017) *Sociology of Families: Change, Continuity, and Diversity*. Thousand Oaks, Ca. and London: Sage.

Cimardi, L. (2015) 'From "tribes" to "regions". Ethnicity and musical identity in western Uganda.' *Perifèria*, revista de recerca i formació en antropologia 20(2). [478–1784–1-PB (2).pdf].

Cindoglu, D. and Unal, D. (2017) 'Gender and sexuality in the authoritarian discursive strategies of "New Turkey".' *European Journal of Women's Studies* 24(1): 39–54.

Clark, L. S. (1998) 'Dating on the net: Teens and the rise of "pure" relationships', pp. 159–183. In S. G. Jones (ed.), *Cybersociety 2.0: Revisiting Computer-Mediated Communication and Community*. Thousand Oaks and London: Sage.

Clarke, L. H. and Griffin, M. (2008) 'Visible and invisible ageing: Beauty work as a response to ageism.' *Ageing & Society* 28(5): 653–674.

Clement, J. (2020) 'The Match Group: Quarterly paid member count 2014–2020.' Statista, 18 August [https://www.statista.com/statistics/449465/paid-dating-subscribers-match-group/].

Clue and Kinsey Institute (2017) 'Technology and sexuality: International Sex Survey', August 9 [https://helloclue.com/articles/sex/technology-modern-sexuality-results-from-clue-kinseys-international-sex-survey].

Cohen (2017), M. T. 'Lying online: The use of deception to find a mate.' *Psychology Today*, December 22 [https://www.psychologytoday.com/us/blog/finding-love-the-scientific-take/201712/lying-online].

Comella, L. (2015) 'Review: Revisiting the feminist sex wars.' *Feminist Studies* 41(2): 437–462.

Commonwealth of Australia (2019) 'Australia's ageing population: Understanding the fiscal impacts over the next decade.' Parliamentary Budget Office, Report No. 02/2019, Canberra [https://www.aph.gov.au/About_Parliament/Parliamentary_Departments/Parliamentary_Budget_Office/Publications/Research_reports/Australias_ageing_population_-_Understanding_the_fiscal_impacts_over_the_next_decade].

Compton, J (2019) ' "The orgasm gap": Why it exists and what women can do about it.' NBC News, 7 April [https://www.nbcnews.com/better/lifestyle/orgasm-gap-why-it-exists-what-women-can-do-about-ncna983311].

Conger, C. (2011) 'Do people lie more on the internet?' *ABC News*, 5 March [https://abcnews.go.com/Technology/people-lie-internet/story?id=13060797].

Conn, E. (2018) 'Gender equality: Argentina's pipe dream.' *The Bubble*, September 14 [https://www.thebubble.com/gender-equality-argentinas-pipe-dream].

Connell, R. (1987) *Gender and Power: Society, the Person and Sexual Politics*. Cambridge: Polity and Blackwell.

Connell, R. and Messerschmidt, J. (2005) 'Hegemonic masculinity: Rethinking the concept.' Gender & Society 19(6): 829–859.

Cooper, D. (2008). 'Golden Oldies discover love at first byte.' ABC News in Science, Friday, 24 October [http://www.abc.net.au/science/articles/2008/10/24/2397187.htm].

Cooper, K., in conjunction with *OkCupid* (2016) 'A digital decade: How sexual opinions and behavior have changed from 2005 to 2015', February 5 [https://theblog.okcupid.com/a-digital-decade-sex-c95e6fb6296b].

Cooper, M. N. (2000) *Disconnected, Disadvantaged and Disenfranchised: Explorations in the Digital Divide*. Washington DC: Consumer Federation of America and Consumers Union Joint Report, October 11.

Couch, D. and Liamputtong, P. (2008) 'Online dating and mating: The use of the internet to meet sexual partners.' *Qualitative Health Research* 18(2): 268–279.

Cousins, S. (2020) 'Contemporary Australia: National identity.' Monash University National Centre for Australian Studies/Open Learning Australia [http://www.abc.net.au/ra/australia/pdf/national_id.pdf].

Croffey, A. (2016) 'Single in Australia. It's not you, it's Aussies.' *Sydney Morning Herald*, February 24 [https://www.smh.com.au/lifestyle/single-its-not-you-its-aussies-20160223-gn11o7.html].

Crossley, P. (2018) 'SSM: A global perspective on Australia's change.' *ABC News*, 11 January [https://www.abc.net.au/news/2017-12-07/same-sex-marriage-ssm-mapped/9234650?nw=0].

Cummings, J. (2020) ' "Now you can see who's around you": Negotiating and regulating gay intimacies on mobile media in the People's Republic of China', pp. 15–30. In J. Cabañes and C. Uy-Tioco (eds), *Mobile Media and Social Intimacies in Asia. Mobile Communication in Asia: Local Insights, Global Implications*. Dordrecht: Springer.

Daily Sabah (2019) 'Internet usage continues to rise in Turkey as 3 in 4 people now online' [https://www.dailysabah.com/turkey/2019/08/27/internet-usage-continues-to-rise-in-turkey-as-3-in-4-people-now-online?gallery_image=undefined#big].

Daneback, K., Månsson, S. A. and Ross, M. (2007) 'Using the internet to find offline sex partners.' *CyberPsychology & Behavior* 10(1): 100–110.

Daniels, J. (2009) 'Rethinking cyberfeminism(s): Race, gender, and embodiment.' *Women's Studies Quarterly* 37(1/2) Spring–Summer: 101–124.

Dating Scout (2020a) 'Victoria Milan review September 2020.' Australian online review site [https://www.datingscout.com.au/victoria-milan/review].

Dating Scout (2020b) 'RSVP review September 2020.' Australian online review site [https://www.datingscout.com.au/rsvp/review#charts].

Dating Scout (2020c) 'Mingle2 review October 2020.' Australian online review site [https://www.datingscout.com.au/mingle2/review#charts].

Davidson, J., Smith, M. M. and Bondi, L. (eds) (2007). *Emotional Geographies*. Aldershot: Ashgate.

Davis, E. M. and Fingerman, K. L. (2016) 'Digital dating: Online profile content of older and younger adults.' *Journal of Gerontology Series B: Psychological Sciences and Social Sciences* 71(6): 959–967.

Debrot, A., Meuwly, N., Muise, A., Impett, E. A. and Schoebi, D. (2017) 'More than just sex: Affection mediates the association between sexual activity and well-being.' *Personality and Social Psychology Bulletin* 43(3): 287|–299.

Debrot, A., Schoebi, D., Perrez, M. and Horn, A. B. (2013) 'Touch as an interpersonal emotion regulation process in couples' daily lives: The mediating role of psychological intimacy.' *Personality and Social Psychology Bulletin* 39(10): 1373–1385.

Deitz, B. (2016) 'Should you delete your dating profile if you met someone on an app? 9 Experts offer their best advice.' *Bustle*, August 25 [https://www.bustle.com/articles/179045-when-should-you-delete-your-dating-profile-if-you-met-on-someone-on-an-app-9].

DeLamater, J. D. and Hyde, J. S. (1998) 'Essentialism vs. social constructionism in the study of human sexuality.' *The Journal of Sex Research* 35(1): 10–18.

DeLamater, J. and Koepsel, E. (2015) 'Relationships and sexual expression in later life: A biopsychological perspective.' *Sexual and Relationship Therapy* 30: 37–59.

DeLamater, J. and Sill, M. (2005) 'Sexual desire in later life.' *Journal of Sex Research* 42(2): 138–149.

Deleuze, G. and Guattari, F. (2004) [1972] *Anti-Oedipus. Capitalism and Schizophrenia Vol. 1*, Translated by Robert Hurley et al. London and New York: Continuum.

Dellisanti, G. (2017) 'The illusion of gender equality.' *The Murmur: Denmark Internationally*, Thursday December 21 [http://murmur.dk/the-illusion-of-gender-equality/#:~:text=Danes%20believe%20equality%20has%20been%20achieved&text=%E2%80%9CIt%20is%20obvious%2C%20also%20to,share%20the%20responsibility%20for%20children].

Delueze, G. and Guattari, F. 1987 [1980] *A Thousand Plateaus: Capitalism and Schizophrenia Vol. 2*, Translated by Brian Massumi. Minneapolis, MN: University of Minnesota Press.

DeMasi, S. (2011) 'Shopping for love: Online dating and the making of a cyber culture of romance', pp 206–214. In S. Seidman, N. Fischer and C. Meeks (eds), *Introducing the New Sexuality Studies*. London and New York: Routledge.

Denmark 2020: Ecommerce Country Report (2020) 'Ecommerce Country Report.' *RetailX*, 1 March [https://retailx.net/product/denmark-2020/].

Denmark F. L. (2002) 'Myths of aging.' *Psi Chi (The National Honor Society in Psychology): Distinguished Lectures/Special Topics* 7(1): 14–21.

Dent, G. (2019) 'Australia slips backwards in global Gender Gap rank. Again.' *Women's Agenda*, December 17 [https://womensagenda.com.au/latest/australia-slips-backwards-in-global-gender-gap-rank-again/].

Deresiewicz, W. (2012) 'A man. A woman. Just friends?' *The New York Times*, April 7 [https://www.nytimes.com/2012/04/08/opinion/sunday/a-man-a-woman-just-friends.html].

Dinh, R., Gildersleve, P., Blex, C. and Yasseri, T. (2018) 'Computational courtship: Understanding the evolution of online dating through large-scale data analysis.' Oxford Internet Institute, arXiv, 26 September [https://arxiv.org/abs/1809.10032].

Dixon, L. (2020) 'How dating apps have gone IRL.' *Australian Financial Review: Weekend Fin*, 7–8, March: 41.

Dodgson, L. (2018) 'Relationship experts say these are the 8 red flags to look out for when you start dating someone—and some are surprisingly common.' *The Independent*, Thursday 25 January [https://www.independent.co.uk/life-style/love-sex/dating-bad-signs-warning-red-flags-new-relationship-love-sex-a8178586.html].

Dolan, E. (2020) 'Online dating: How markets and demographics differ', February 11 [https://blog.globalwebindex.com/chart-of-the-week/differences-among-online-daters/].

Donnelly, D. and Burgess, E. (2008) 'The decision to remain in an involuntarily celibate relationship.' *Journal of Marriage and Family* 70: 519–535.

Döring, N. (2009) 'The internet's impact on sexuality: A critical review of 15 years of research.' *Computers in Human Behaviour* 25(5): 1089–1101.

Dowling, A. (2014) '9 Things you didn't know about dating for seniors.' Stitch: The Stitch Blog [https://www.stitch.net/blog/2014/05/9-things-you-didnt-know-dating-for-seniors/].

Drummond. M. J. (2010) 'The natural: An autoethnography of a masculinized body in sport.' *Men & Masculinities* 12(3): 374–389.

Drummond. M. J. (2002) 'Sport and images of masculinity: The meaning of relationships in the life course of "elite" male athletes.' *The Journal of Men's Studies* 10(2): 129–141.

Duerringer. C. (2015) 'Be a man: Buy a car! Articulating masculinity with consumerism in *Man's Last Stand*.' *Southern Communication Journal* 80(2): 137–152.

Dwyer, G. (2016) 'In the race to get married before turning 30, many young Indian-Australians are returning to their homeland and actively seeking their parents' help in search of the perfect partner.' *SBS*, 12 April [https://www.sbs.com.au/programs/untold-australia/article/2016/04/12/meet-young-indian-australians-returning-india-search-love?cid=inbody%3Aethnic-dating-in-australia-what-are-the-options].

EAC (EveryAgeCounts) (no date) 'Does ageism exist in all cultures?' [https://www.everyagecounts.org.au/does_ageism_exist_in_all_cultures].

Eastwick, P. and Finkel, E. (2008) 'Sex differences in mate preferences revisited: Do people know what they initially desire in a romantic partner?' *Journal of Personality and Social Psychology* 94(2): 245–264.

Eisenlauer, V. and Hoffmann, C. R. (2010) 'Once upon a blog . . . Storytelling in weblogs', pp 79–108. In C. R. Hoffman (ed.), *Narrative Revisited: Telling a Story in the Age of New Media*. Amsterdam: John Benjamins.

Elbow, P. (1993) 'The uses of binary thinking.' *Journal of Advanced Composition* 14: 22–51.

Eleuteri, S., Tripodi, F., Petruccelli, I., Rossi, R. and Simonelli, C. (2014) 'Questionnaires and scales for the evaluation of the online sexual activities: A review of 20 years of research.' *Cyberpsychology: Journal of Psychosocial Research on Cyberspace* 8(1). DOI: 10.5817/CP2014–1-2.

Elias, N. [1939] (2000) *The Civilizing Process: Sociogenetic and Psychogenetic Investigations*. Oxford: Blackwell.

Elias, N. and Dunning, E. (1986) *Quest for Excitement: Sport and Leisure in the Civilizing Process*. Oxford: Basil Blackwell.

Elin, A. (2014) 'Matchmakers help those over 60 handle dating's risks and rewards.' *The New York Times*, March 29 [http://www.nytimes.com/2014/03/29/your-money/matchmakers-can-help-those-over-60-navigate-datings-risks-and-rewards.html].

EliteSingles Editor (no date) 'Older dating: It's never too late to find love.' *EliteSingles Magazine* [https://www.elitesingles.com.au/em/online-dating/older-dating].

Ellison, N. B., Hancock, J. T. and Toma, C. L. (2012) 'Profile as promise: A framework for conceptualizing veracity in online dating self-presentations.' *New Media & Society* 14(1): 45–62.

Ellison, N., Heino, R. and Gibbs, J. (2006) 'Managing impressions online: Self-presentation processes in the online dating environment.' *Journal of Computer-Mediated Communication* 11(2): 415–441.

England, P. and Bearak, J. (2014) 'The sexual double standard and gender differences in attitudes toward casual sex among US university students.' *Demographic Research* 30: 1327.

Equal Measures2030 (2019) 'Harnessing the power of data for gender equality: Introducing the 2019 EM2030 SDG Gender Index' [https://data.em2030.org/wp-content/uploads/2019/05/EM2030_2019_Global_Report_ENG.pdf].

Erasmus (2017) 'Why secular Denmark is using a sledgehammer to protect the sacred: An odd blasphemy case in Denmark.' *The Economist*, February 26 [https://www.economist.com/erasmus/2017/02/26/why-secular-denmark-is-using-a-sledgehammer-to-protect-the-sacred].

Erjavec, K. and Fišer, S. Z. (2016) 'Aging adults about online dating: "I am back on the relationship market".' *Polish Sociological Review* 195: 361–371.

Escasa-Dorne, M. and Jankowiak, W. (2018) 'Do women really desire casual sex? Analysis of a popular adult online dating/liaison site', pp. 369–395. In W. Shapiro (ed.), *Focality and Extension in Kinship: Essays in Memory of Harold W. Scheffler*. Canberra: ANU Press.

European Commission on Denmark (2020) 'Commission staff working document, Country Report Denmark 2020', 26 February. Communication from the Commission to the European Parliament, The European Council, The Council, The European Central Bank and The Eurogroup [https://ec.europa.eu/info/sites/info/files/2020-european_semester_country-report-denmark_en.pdf].

Evans, M., Haussegger, V., Halupka, M. and Rowe, P. (2018) 'From girls to men: Social attitudes to gender equality in Australia.' Canberra, Australia: 50/50 by 2030 Foundation, Institute for Governance and Policy Analysis/University of Canberra.

Evans, M., Jennings, W. and Stoker, G. (2020) 'Democracy2025—How does Australia compare: What makes a leading democracy?' Report No. 6 [https://www.democracy2025.gov.au/documents/Democracy2025-report6.pdf].

Evans, S. J. (2020) 'Not so 'appy.' *The Sun*, 30 March [https://www.thesun.co.uk/fabulous/11141302/coronavirus-lockdown-dating-app-horror-stories/].

Evason, N. (2016) 'Australian culture.' *Cultural Atlas, SBS* [https://culturalatlas.sbs.com.au/australian-culture/australian-culture-core-concepts].

Evason, N. (2015) 'Chinese culture.' *Cultural Atlas* [culturalatlas.sbs.com.au/chinese-culture/chinese-culture-core-concepts#chinese-culture-core-concepts].

Fakoya, O. A., McCorry, N. K. and Donnelly, M. (2020) 'Loneliness and social isolation interventions for older adults: A scoping review of reviews.' *BMC Public Health* 20(1), February 14 [https://pubmed.ncbi.nlm.nih.gov/32054474/].

Falk, P. and Campbell, C. (eds) (1997) *The Shopping Experience*. London: Sage.

Farah (2015) 'Surviving machismo culture in Argentina.' *GO! Girl Guides*, Solo Female Travel, November 2 [https://gogirlguides.com/solo-female-travel/surviving-machismo-culture-in-argentina/].

Fahs, B., Swank, E. and Shambe, A. (2020) '"I just go with it": Negotiating sexual desire discrepancies for women in partnered relationships.' *Sex Roles* 83: 226–239.

Faulkner (2000) 'Dualisms, hierarchies and gender in engineering.' *Social Studies of Science* 30(5): 759–792.

Featherstone, M. and Hepworth, M. (1991) 'The mask of ageing and the postmodern life course', pp. 371–389. In M. Featherstone, M. Hepworth and B. Turner (eds), *The Body: Social Process and Cultural Theory*. London: Sage.

Feeney, C. (2010) 'On the hunt for Princess Charming.' *The Sydney Morning Herald*, June 30 [https://www.smh.com.au/lifestyle/life-and-relationships/on-the-hunt-for-princess-charming-20100630-zjpq.html].

Fehr, C. (2011) 'Feminist philosophy of biology.' *Stanford Encyclopedia of Philosophy*, June 22 [https://plato.stanford.edu/entries/feminist-philosophy-biology/].

Feliciano, C., Robnett, B. and Komaie, G. (2009) 'Gendered racial exclusion among white internet daters.' *Social Science Research* 38(1), March: 39–54.

Ferdman, R. A. (2016) 'How well online dating works, according to someone who has been studying it for years.' *The Washington Post*, March 23 [https://www.washingtonpost.com/news/wonk/wp/2016/03/23/the-truth-about-online-dating-according-to-someone-who-has-been-studying-it-for-years/].

Ferguson, L. (2011) 'The campaign to end loneliness.' *Working with Older People* 15(2): 66–70.

Field, T. (2018) 'Online dating profiles and problems in older adults: A review.' *OBM Geriatrics* 2(3): doi:10.21926/obm.geriatr.1803012 [http://www.lidsen.com/journals/geriatrics/geriatrics-02-03-012].

Fileborn, B., Lyons, A., Hinchliff, S., Brown, G., Heywood, W., Dow, B., Malta, S. and Minichiello, V. (2017) 'Improving the sexual lives of older Australians: Perspectives from a qualitative study.' *Australasian Journal of Ageing* 36(4): E36–E42.

Fileborn, B., Thorpe, R., Hawkes, G., Minichiello, V. and Pitts, M. (2015) 'Sex and the (older) single girl: Experiences of sex and dating in later life.' *Journal of Aging Studies* 33: 67–75.

Finkel, E., Eastwick, P., Karney, B., Reis, H. and Sprecher, S. (2012) 'Online dating: A critical analysis from the perspective of pyschological science.' *Psychological Science in the Public Interest* 13(1): 3–66.

Fiore, A. T., Taylor, L. S., Zhong, X., Mendelsohn, G. A. and Cheshire, C. (2010). 'Who's right and who writes: People, profiles, contacts, and replies in online dating', pp. 1–10. 43rd Hawaii International Conference on System Sciences. IEEE.

Fischer, C. S. (1994) *America Calling: A Social History of the Telephone to 1940*. Berkeley, CA: University of California Press.

Fisman, R., Iyengar, S., Kamenica, E. and Simonson, I. (2006) 'Gender differences in mate selection: Evidence from a speed-dating experiment.' *Quarterly Journal of Economics* 121: 673–697.

Fitzclarence, L. and Hickey, C. (2001) 'Real footballers don't eat quiche: Old narratives in new times.' *Men & Masculinities* 4(2): 118–139.

Fleming, P. (2015) 'The art of flirtation: Simmel's coquetry without end', pp 19–30. In D. Hoffman-Schwartz, B. N. Nagel, and L. S. Stone (eds), *Flirtations: Rhetoric and Aesthetics This Side of Seduction*. New York: Fordham University Press.

Flynn, S. (2019) 'Rethinking debates in narrative methods: Narrative orthodoxy and research challenges with children with intellectual disability.' *Scandinavian Journal of Disability Research* 21(1): 167–176.

Fong, E., Wellman, B., Kew, M. and Wilkes, R. (2001) 'Correlates of the digital divide: Individual, household, and spatial variation.' Report to Office of Learning Technologies, June. Human Resources and Development, Canada.

Ford, L. (2019) 'Not one single country set to achieve gender equality by 2030.' *The Guardian*, Monday 3 June [https://www.theguardian.com/global-development/2019/jun/03/not-one-single-country-set-to-achieve-gender-equality-by-2030].

Ford Foundation (2005) *Sexuality and Social Change: Making the Connection—Strategies for Action and Investment*. New York: Ford Foundation.

Foreign Correspondent (2020) 'All the Single Men.' *ABC News*, 18 January [https://www.abc.net.au/foreign/all-the-single-men/123622].

Fottrell, Q. (2017) 'Most young people don't use Tinder for sex or love—here's why they really use it.' *MarketWatch*, April 8 [https://www.marketwatch.com/story/most-people-dont-use-tinder-for-love-or-sex-heres-why-they-really-use-it-2017–03–23?mod=article_inline].

Foucault, M. (1990) *The History of Sexuality: Volume One, An Introduction*. London: Penguin.

Foucault, M. (1980) *Power/Knowledge: Selected Interviews and Other Writings 1972–1977*. London: Harvester Press.

Foucault, M. (1977) 'Preface', to G. Deleuze and F. Guattari (2004) [1972] *Anti-Oedipus. Capitalism and Schizophrenia Vol. 1*, Translated by Robert Hurley et al. London and New York: Continuum.

Freedom House—Australia (2020a) 'Freedom in the World 2020: Australia' [https://freedomhouse.org/country/australia/freedom-world/2020].

Freedom House—Denmark (2020b) 'Denmark' [https://freedomhouse.org/country/denmark/freedom-world/2020].

Freedom House—UK (2020c) 'Freedom in the World 2020: United Kingdom [https://freedomhouse.org/country/united-kingdom/freedom-world/2020].

Freedom House (2019ba) 'Freedom on the Net: Australia 2019' [https://freedomhouse.org/country/australia/freedom-net/2019].

Freedom House (2019b) 'Freedom on the Net: Uganda 2019' [https://freedomhouse.org/country/uganda/freedom-net/2019].

Freedom House (2018) 'Country Data—Countries and Territories' [https://freedomhouse.org/countries/freedom-world/scores].

Frohlick, S. and Migliardi, P. (2011) 'Heterosexual profiling: Online dating and "becoming" heterosexualities for women aged thirty and older in the digital era.' *Australian Feminist Studies* 26: 73–88.

Furlong, M. S. (1989) 'An online community for older adults: The SeniorNet network.' *Journal of Communication* 39(3): 145–153.

Furtado, D. (2015) 'Ethnic intermarriage', pp. 118–122. In *International Encyclopedia of the Social and Behavioural Sciences,* 2nd edition. Elsevier.

Gagnon, J. H., and Simon, W. (2005) [1973] *Sexual Conduct: The Social Sources of Human Sexuality*. New Brunswick NJ: Aldine Transaction.

Garcia, J., Reiber, C., Massey, S. and Merriwether, A. (2012) 'Sexual hookup culture: A review.' *Review of General Psychology* 16(2): 161–176.

Garcia-Favaro, L. (2016) 'Porn trouble: On the sexual regime and travels of postfeminist biologism.' *Australian Feminist Studies* 30(86): 366–376.

Garde-Hansen, J. and Gorton, K. (2013) *Emotion Online: Theorizing Affect on the Internet*. Basingstoke: Palgrave Macmillan.

Garfinkel, H. (1964) 'Studies of the routine grounds of everyday activities.' *Social Problems* 11(3): 225–250.

Gatter, K. and Hodkinson, K. (2016) 'On the differences between Tinder versus online dating agencies: Questioning a myth. An exploratory study.' *Cogent Psychology* 3(1) [https://doi.org/10.1080/23311908.2016.1162414].

Geller, L. (2017) 'Twice as many LGBTQ+ singles use dating apps as heterosexual ones. Here's why that matters.' *A Plus*, August 21 [https://articles.aplus.com/a/how-dating-apps-benefit-lgbtq-singles?no_monetization=true].

Gewirtz-Meydan A. and Ayalon, L. (2018) 'Forever young: Visual representations of gender and age in online dating sites for older adults.' *Journal of Women & Aging* 30(6): 484–502.

Gewirtz-Meydan, A., Hafford-Letchfield, T., Benyamini, Y., Phelan, A., Jackson, J. and Ayalon, L. (2018) 'Ageism and sexuality', pp149–162. In L. Ayalon and C. Tesch-Römer (eds), *Contemporary Perspectives on Ageism. International Perspectives on Aging, vol 19*. Cham: Springer.

Giakoumelos, P. (2017) 'For those Australians who are looking for love within their own ethnic group, there's never been more variety in the diverse range of dating apps and matchmaking services available.' *SBS World News Radio*, 20 February [https://www.sbs.com.au/radio/article/2017/02/20/ethnic-dating-australia-what-are-options].

Gibbs, J. L., Ellison, N. B. and Heino, R. D. (2006) 'Self-presentation in online personals: The role of anticipated future interaction, self-disclosure, and perceived success in internet dating.' *Communication Research* 33(2): 152–177.

Giddens, A. (1992) *The Transformation of Intimacy: Sexuality, Love and Eroticism in Modern Societies*. Cambridge: Blackwell/Polity Press.

Giddens, A. (1990) *The Consequences of Modernity*. Palo Alto, CA: Stanford University Press.
Gilbert, S. M. and Gubar, S. (1980) *The Madwoman in the Attic: The Woman Writer and the Nineteenth-Century Literary Imagination*. New Haven, CT: Yale University Press.
Gillies, V. (2003) 'Family and intimate relationships: A review of the sociological research, families and social capital.' ESRC Research Group Working Paper No. 2, June, South Bank University, London.
Gilmore, P. (2019) 'Casual sex and how to have positive hookups, according to 5 women.' *Cosmopolitan*, September 23 [https://www.cosmopolitan.com/uk/love-sex/sex/a29145757/casual-sex/].
Girls not Brides (no date) [https://www.girlsnotbrides.org/child-marriage/uganda/].
Glick, E. (2000) 'Sex positive: Feminism, queer theory, and the politics of transgression.' *Feminist Review* 64: 19–45.
Goffman, E. (1987) *Gender Advertisements*. New York: Harper and Row.
Goffman, E. (1968) *Stigma: Notes on the Management of Spoiled Identity*. Harmondsworth: Penguin.
Gong C. and Kendig, H. (2016) 'Ageing and social change in Australia', pp. 19–46. In H. Kendig, P. McDonald, and J. Piggott (eds), *Population Ageing and Australia's Future*. Canberra: ANU Press.
Goñi, U. (2016) 'Argentinian activists pin blame on machismo as attacks on women rise.' *The Guardian*, Thursday 27 October [https://www.theguardian.com/world/2016/oct/27/argentinian-activists-machismo-attacks-women-niunamenos].
Granwal, L. (2019) 'Internet usage in Australia: Statistics & facts.' Statista, August 23 [https://www.statista.com/topics/5261/internet-usage-in-australia/].
Green, A. I. (2013) ' "Erotic capital" and the power of desirability: Why "honey money" is a bad collective strategy for remedying gender inequality.' *Sexualities* 16(1–2): 137–158.
Green, A. I. (2007) 'Queer theory and sociology: Locating the subject and the self in sexuality studies.' *Sociological Theory* 25: 26–45.
Green, L. (2014) 'Why millions of Muslims are signing up for online dating.' *BBC News, Magazine*, 10 December [https://www.hurriyetdailynews.com/mothers-only-career-should-be-motherhood-turkish-health-minister-says-76360].
Griffiths, M. (2018) 'Baby boomers re-entering dating game more vulnerable to sexually transmitted infections.' ABC News, Thursday 18 January [https://www.abc.net.au/news/2018-01-18/baby-boomers-back-in-the-dating-game-not-using-condoms-research/9337160].
Groves, J. M. and Chan, A. H. (2017) 'Love in the time of "settling": Forbidden knowledge and modern singles advice.' *Journal of Sociology* 54(1): 141–155.
Guo, D. (2017) 'Well educated, single women in their late 20s are called "leftover women" in China.' *Quartz*, July 4 [https://qz.com/1019300/in-china-single-women-are-looked-down-upon-regardless-of-their-success/].
Gupta, K. (2015) 'Compulsory sexuality: Evaluating an emerging concept.' *Signs* 41(1): 131–154.
Hague, G., Thiara, R. K. and Turner, A. (2011) 'Bride-price and its links to domestic violence and poverty in Uganda: A participatory action research study.' *Women's Studies International Forum* 34(6): 550–561.
Hakim, C. (2012) *The New Rules: Internet Dating, Playfairs and Erotic Power*. London: Gibson Square.

Hamilton N. F. (2016) 'Romantic relationships and online dating', pp. 144–160. In A. Attrill and C. Fullwood (eds), *Applied Cyberpsychology*. London: Palgrave Macmillan.

Haraway, D. (1991) 'A cyborg manifesto: Science, technology and social feminism in the late twentieth century', pp 149–181. In *Simians, Cyborgs and Women: The Reinvention of Nature*. New York: Routledge.

Hardey, M. (2008) 'The formation of social rules for digital interactions.' *Information, Communication & Society* 11(8): 1111–1131.

Hardey, M. (2002) 'Life beyond the screen: Embodiment and identity through the internet.' *Sociological Review* 50(4): 570–585.

Harper, H. (2018) 'Finding love: The etiquette of online dating.' *Life Matters*, Tuesday 27 February. Radio National, Australian Broadcasting Commission podcast [https://www.abc.net.au/radionational/programs/lifematters/finding-love:-the-etiquette-of-online-dating/9485984].

Harris, S. (no date) 'How to meet Ugandan women (dating in Kampala).' *Global Seducer* [https://www.globalseducer.com/ugandan-women/].

Harrison, K. (2013) *Sexual Deceit: The Ethics of Passing*. Lanham, Maryland: Lexington Books.

Hayfield, N. and Clarke, V. (2012) '"I'd be just as happy with a cup of tea": Women's accounts of sex and affection in long-term heterosexual relationships.' *Women's Studies International Forum* 35(2): 67–74.

Haywood C. (2018) 'Mobile romance: Tinder and the navigation of masculinity', pp. 131–166. In *Men, Masculinity and Contemporary Dating*. London: Palgrave Macmillan.

Hearn, J. (2007) 'Men, ageing and power: Can men's ageing challenge patriarchy?' Östersund: Mittuniversitetet (Mid Sweden University) [http://www.diva-portal.org/smash/get/diva2:1653/FULLTEXT01.pdf#page=73].

Hearn, J. and Sandberg, L. (2009) 'Older men, ageing and power: Masculinities theory and alternative spatialised theoretical perspectives.' *Sextant: revue du Groupe interdisciplinaire d'etudes sur les femmes et le genre*, 27: 147–163.

Hearn, J., Nordberg, M., Andersson, K., Balkmar, D., Gottzén, L., Klinth, R., Pringle, K. and Sandberg, L. (2012) 'Hegemonic masculinity and beyond: 40 years of research in Sweden.' *Men and Masculinities* 15(1): 31–55.

Hedetoft, U. (2010) 'Denmark versus multiculturalism', pp. 111–129. In S. Vertovec and S. Wessendorf (eds), *The Multiculturalism Backlash: European Discourses, Policies and Practices*. London: Routledge.

Hedetoft, U. (2006) 'Denmark: Integrating immigrants into a homogenous welfare state.' *Migration Policy Institute*, Migration Information Source, November 1 [https://www.migrationpolicy.org/article/denmark-integrating-immigrants-homogeneous-welfare-state].

Heino, R., Ellison, N. and Gibbs, J. (2010) 'Relationshopping: Investigating the market metaphor in online dating.' *Journal of Social and Personal Relationships*, 27: 427–447.

Helsper, E. J. and Whitty, M. (2010) 'Netiquette within married couples: Agreement about acceptable online behavior and surveillance between partners.' *Computers in Human Behaviour* 26(5): 916–926.

Hemmings, C. (2005) 'Invoking affect: Cultural theory and the ontological turn.' *Cultural Studies* 19(5): 548–567.

Hendricks, S. (2018) 'Here's what dating is like in 20 countries around the world.' *Insider*, May 30 [insider.com/what-dating-is-like-in-different-countries-2018–5].

Henley, J. (2019) 'Breaking up is harder to do in Denmark after divorce law changes.' *The Guardian*, Monday 22 July [https://www.theguardian.com/world/2019/jul/22/breaking-up-is-harder-to-do-in-denmark-after-divorce-law-changes].

Henriksson, A. (2019) 'Singles' activities: Sociability and the ambiguities of singledom.' *Families, Relationships and Societies* 8(1): 37–52.

Henriksson, A. (2014) 'Organising intimacy—Exploring heterosexual singledoms at Swedish singles activities.' Unpublished PhD Dissertation, Sociology. Karlstad, Sweden: Karlstad University.

Henriques-Gomes, L. (2020) 'Rosie Batty "dismayed" by decision to given Bettina Arndt an Australia Day honour.' *The Guardian*, Monday 27 January [https://www.theguardian.com/australia-news/2020/jan/27/rosie-batty-dismayed-by-decision-to-give-bettina-arndt-an-australia-day-honour#img-1].

Hepworth, M. (1995) 'Positive ageing: What is the message?' In S. Nettleton, R. Brunton and R. Burrows (eds), *The Sociology of Health Promotion: Critical Analyses of Consumption, Lifestyle and Risk*. London: Routledge.

Herbenick, M. (2014) 'The influence of attraction to partner on heterosexual women's sexual and relationship satisfaction in long-term relationships.' *Archives of Sexual Behaviour* 43(3): 563–570.

Hergovich, P. and Ortega, J. (2018) 'The strength of absent ties: Social integration via online dating', September 14. *SSRN* [https://ssrn.com/abstract=3044766 or http://dx.doi.org/10.2139/ssrn.3044766].

Hess, A. and Flores, C. (2018). 'Simply more than swiping left: A critical analysis of toxic masculine performances on Tinder Nightmares.' *New Media & Society* 20: 1085–1102.

Hickok, H. (2017) 'What to do when you're dating a guy with problems below the belt.' StyleCaster, blog, October 30 [https://stylecaster.com/dating-man-with-erectile-dysfunction/].

Hill, F. (2020) 'What it's like to date after middle age.' *The Atlantic*, January 8 [https://www.theatlantic.com/family/archive/2020/01/dating-after-middle-age-older/604588/].

Hillier, L. and Harrison, L. (2007) 'Building realities less limited than their own: Young people practising same-sex attraction on the internet.' *Sexualities* 10(1): 82–100.

Hillier, L., Mitchell, K. J. and Ybarra, M. L. (2012) 'The internet as a safety net: Findings from a series of online focus groups with LGB and non-LGB young people in the United States.' *Journal of LGBT Youth* 9(3): 225–246.

Hirsch, J. B. (2018) 'Wild country: Australian masculinity from the frontier to the social front.' Research Paper, Bachelor of Visual Arts (Honours), October. Sydney, Australia: Sydney College of the Arts, University of Sydney.

Hirst, J. (2004) 'The distinctiveness of Australian democracy.' Paper presented as a lecture in the Senate Occasional Lecture Series at Parliament House, Canberra, 10 September [https://www.aph.gov.au/binaries/senate/pubs/pops/pop42/hirst.pdf].

Hitsch, G., Hortaçsu, A. and Ariely, D. (2010) 'What makes you click? Mate preferences in online dating.' *Quantitative Marketing and Economics* 8(4): 393–427.

Hochschild, A. R. (2003) *The Commercialization of Intimate Life: Notes from Home and Work*. Berkeley, CA: University of California Press.

Hochschild, A. R. (1983) *The Managed Heart: Commercialization of Human Feeling*. Berkeley, CA: University of California Press.

Hockey, M. (2020) 'Take 5: The five best dating apps for over 50s.' *MSN*, 9 April [https://www.msn.com/en-au/lifestyle/familyandrelationships/the-five-best-dating-apps-for-over-50s/ar-BB12m0AZ?li=AA2FZ8I].

Hoffman-Schwartz, D. (2015) 'Barely covered banter: Flirtation in "Double Indemnity"', pp. 13–18. In D. Hoffman-Schwartz, B.N. Nagel and L. S. Stone (eds), *Flirtations: Rhetoric and Aesthetics This Side of Seduction*. New York: Fordham University Press.

Hoffman-Schwartz, D., Nagel, B. N. and Stone, L. S. (2015) '"Almost nothing; almost everything": An introduction to the discourse of flirtation', pp 1–10. In D. Hoffman-Schwartz, B. N. Nagel and L. S. Stone (eds), *Flirtations: Rhetoric and Aesthetics This Side of Seduction*. New York: Fordham University Press.

Hogan, B., Dutton, W. and Li, N. (2011) *A Global Shift in the Social Relationships of Networked Individuals: Meeting and Dating Online Comes of Age*. Oxford Internet Institute, University of Oxford. [https://www.oii.ox.ac.uk/archive/downloads/publications/Me-MySpouse_GlobalReport.pdf].

Hogeboom, D., McDermott, R. J., Perrin, K. M., Osman, H. and Bell-Ellison, B. A. (2010) 'Internet use and social networking among middle aged and older adults.' *Educational Gerontology* 36(2): 93–111.

Hollway, W. (1989) *Subjectivity and Method in Psychology: Gender, Meaning and Science*. London: Sage.

Holmes, K. and Pinto, S. (2013) 'Gender and sexuality', pp. 308–331. In A. Bashford and S. Macintryre (eds), *The Cambridge History of Australia*. Melbourne, Victoria: Cambridge University Press.

Holmes, M. (2015a) 'Researching emotional reflexivity.' *Emotion Review* 7(1), January: 61–66.

Holmes, M. (2015b) 'Men's emotions: Heteromasculinity, emotional reflexivity, and intimate relationships.' *Men and Masculinities* 18(2): 176–192.

Holmes, M. (2011) 'Emotional reflexivity in contemporary friendships: Understanding it using Elias and Facebook etiquette.' *Sociological Research Online* [http://www.socresonline.org.uk/16/1/11.html].

Holmes, M. (2010) 'The emotionalization of reflexivity.' *Sociology* 4(1): 139–154.

Holohan, M. (2016) 'OkCupid report shows how many dates couples usually go on before having sex.' *Today*, February 24 [https://www.today.com/health/okcupid-report-shows-how-many-dates-couples-usually-go-having-t75146].

Holt-Lunstad, J., Robles, T. and Sbarra, D. (2017) 'Advancing social connection as a public health priority in the United States.' *American Psychologist* 72(6): 517–530.

Holt-Lunstad J., Smith T. B. and Layton, J. B. (2010) 'Social relationships and mortality risk: A meta-analytic review.' *PLoS Med* 7(7): e1000316. DOI: 10.1371/journal.pmed.100031.

Holtug, N. (2013) 'Danish multiculturalism, where art thou?', pp. 190–215. In R. Taras (ed.), *Challenging Multiculturalism: European Models of Diversity*. Edinburgh: Edinburgh University Press.

Hong, W. Z. and Lia, C. C. (2019) 'Trends and research issues of online dating: A review of academic publications from 2005 to 2019.' *Cognitive Cities: Second International Conference, IC3*, Revised Selected Papers. September 3–6, Kyoto, Japan.

Hookway, N. (2008) '"Entering the blogosphere": Some strategies for using blogs in social research.' *Qualitative Research* 8(1): 91–113.

Hsieh, H-F. and Shannon, S. (2005) 'Three approaches to qualitative content analysis.' *Qualitative Health Research*, 15(9): 1277–1288.

Hu, J. X. and Nash, S. T. (2019) *Marriage and the Family: Mirror of a Diverse Global Society*. Abingdon, Oxon and New York: Routledge.

Hughes, J. (2011) 'Are older men taking sexuality as prescribed?' *Australian Feminist Studies* 26(67): 89–102.

Hürriyet daily news (2019) 'Turks have negative view on online dating: Study', August 17 [https://www.hurriyetdailynews.com/turks-have-negative-view-on-online-dating-study-145835].

Hürriyet daily news (2015) 'Mothers' only career should be motherhood, Turkish Health Minister says', January 1 [https://www.hurriyetdailynews.com/mothers-only-career-should-be-motherhood-turkish-health-minister-says-76360].

Huyck, M. H. (2020) 'Falling in love in later life', pp. 177–191. In A. Abela, S. Vella and S. Piscopo (eds), *Couple Relationships in a Global Context: Understanding Love and Intimacy Across Cultures*. European Family Therapy Association Series. Cham, Switzerland: Springer.

Illouz, E. (2013) *Why Love Hurts: A Sociological Explanation*. Cambridge, UK: Polity.

Illouz, E. (2009) 'Emotions, imagination and consumption: A new research agenda.' *Journal of Consumer Culture* 9(3): 377–413.

Illouz, E. (2006) 'Réseaux amoureux sur internet.' *Réseaux*, 4: 269–272.

Inglehart, R., Haerpfer, C., Moreno, A., Welzel, C., Kizilova, K., Diez-Medrano, J., Lagos, M., Norris, P., Ponarin, E. and Puraneneds, B. (2014) 'World values survey: Round six—country-pooled datafile', [https://www.worldvaluessurvey.org/WVS DocumentationWV6.jsp. Madrid: JD Systems Institute].

Inglehart, R. and Welzel, C. (2010) 'Changing mass priorities: The link between modernization and democracy.' *Perspectives on Politics* 8(2), June: 551–567 (see World Values Survey Cultural Map 2005–2008).

Internet live stats (2016) 'Uganda internet users' [https://www.internetlivestats.com/internet-users/uganda/].

Internetworldstats (2020) 'Internet penetration in Africa: 2020–Q1–March.' Internet World Stats: Usage and Population Statistics [https://www.internetworldstats.com/stats1.htm].

Iqbal, M. (2020) 'Tinder revenue and usage statistics (2020).' *Business of Apps*, March 25 [https://www.businessofapps.com/data/tinder-statistics/].

Island, H. (2018) 'Netiquette'. In B. Warf (ed.), *The SAGE Encylopedia of the Internet* [http://sk.sagepub.com/reference/the-sage-encyclopedia-of-the-internet-3v/i7242.xml].

ITU (2020) 'The ICT Development Index (IDI): Conceptual framework and methodology' [https://www.itu.int/en/ITU-D/Statistics/Pages/publications/mis/methodology.aspx].

ITU-IDI (2017) 'ICT Development Index 2017.' [https://www.itu.int/net4/ITU-D/idi/2017/index.html].

Ivana, G-I. (2018) *Social Ties in Online Networking*. London. Palgrave Macmillan.

Iversen, H. R. (2006) 'Secular religion and religious secularism: A profile of the religious development in Denmark since 1968.' *Nordic Journal of Religion and Society* 19(2): 75–92.

Jackson, M. (1984) 'Sex research and the construction of sexuality: A tool of male supremacy?' *Women's Studies International Forum* 7(1): 43–51.

Jackson, S. (1999) *Heterosexuality in Question.* London: Sage.

Jackson, S. and Scott, S. (2004) 'Sexual antinomies in late modernity.' *Sexualities* 7(2): 233–248.

Jagger, E. (2005) 'Is thirty the new sixty? Dating, age and gender in a postmodern, consumer society.' *Sociology* 39(1): 89–106.

Jagger, E. (2001) 'Marketing Molly and Melville: Dating in a postmodern, consumer society.' *Sociology* 35(1): 39–57.

Jamieson, L. (2013) 'Personal relationships, intimacy and the self in a mediated and global digital age', pp. 13–33. In K. Orton-Johnson and N. Prior (eds), *Digital Sociology: Critical Perspectives*. Basingstoke: Palgrave Macmillan.

Jamieson, L. (1999) 'Intimacy transformed? A critical look at the "pure relationship".' *Sociology* 33(3): 477–494.

Jamieson, L. and Simpson, R. (2013) *Living Alone: Globalization, Identity and Belonging*. Basingstoke: Palgrave Macmillan.

Jane, E. A. (2014) '"Your a Ugly, Whorish, Slut": Understanding E-bile.' *Feminist Media Studies* 14(4): 531–546.

Johnson, J. (2020) 'Household internet penetration in the United Kingdom (UK) 1998–2019.' Statista, February 14 [https://www.statista.com/statistics/275999/household-internet-penetration-in-great-britain/].

Johnson, J. (2017) 'Dating_MissRepresentation.Com: Black women's lived love-hate relationship with online dating.' Unpublished PhD thesis. Dissertations 1363, Southern Illinois University Carbondale [http://opensiuc.lib.siu.edu/dissertations/1363].

Jones, B. T. (2017) 'What was the White Australia Policy, and how does it still affect us now?' *The Conversation*, NITV, SBS, 10 April [https://www.sbs.com.au/nitv/nitv-news/article/2017/04/10/what-was-white-australia-policy-and-how-does-it-still-affect-us-now].

Jones, H., Kelley, N. and Rantenen, K. (2019) 'Women and work', pp. 82–111. In J. Curtice, E. Clery, J. Perry, M. Phillips and N. Rahim (eds), *British Social Attitudes: The 36th Report*. London: The National Centre for Social Research.

Jørgensen, M. J. (2011) 'Identifying key topics for the description of sexual behaviour in the Danish population: A qualitative study.' Conference poster, The 19th ISSTDR Conference, International Society for STD Research [https://www.forskningsdatabasen.dk/en/catalog/2389232119].

Jotanovic, D. (2019) 'Australian society is obsessed with masculinity—and it's making me tired.' *The Guardian*, Saturday 24 August [https://www.theguardian.com/commentisfree/2019/aug/24/australian-society-is-obsessed-with-masculinity-and-its-making-me-tired].

Juhasz, S. (1998) 'Lesbian romance fiction and the plotting of desire: Narrative theory, lesbian identity, and reading practice.' *Tulsa Studies in Women's Lit*erature 17(1): 65–82.

Jung, E., Walden, G., Johnson, A. and Sundary, S. S. (2017) 'Social networking in the aging context: Why older adults use or avoid Facebook.' *Telematics & informatics* 34(7): 1071–1080.

Jupp, J. (2007) *From White Australia to Woomera: The Story of Australian Immigration*, 2nd edition. Melbourne: Cambridge University Press.

Kanaabi, M. (2020) '42% of Uganda's population now connected to internet.' *New Vision*, Friday, August 28 [https://www.newvision.co.ug/news/1517233/-uganda-population-connected internet#:~:text=A%20report%20by%20online%20retailer,stands%20at%2044.5%20million%20people].

Karantzas, G. (2018) 'Mind the gap—does age difference in relationships matter?' *The Conversation*, April 20 [https://theconversation.com/mind-the-gap-does-age-difference-in-relationships-matter-94132].

Kavanagh, M. (2019) 'Household food spending divides the world.' *Financial Times*, Special Report—Sustainable Food and Agriculture, January 8 [https://www.ft.com/content/cdd62792-0e85-11e9-acdc-4d9976f1533b].

Kelley, J. M. and Malouf, R. (2013) 'Blind dates and mate preferences: An analysis of newspaper matches.' *Evolutionary Psychology* 11(1): 1–8.

Kemp, S. (2020a) 'Digital 2020: Australia.' *Datareportal*, 18 February [https://datareportal.com/reports/digital-2020-australia].

Kemp, S. (2020b) 'Digital 2020: Turkey.' *Datareportal*, 18 February [https://datareportal.com/reports/digital-2020-turkey#:~:text=Internet%20penetration%20in%20Turkey%20stood%20at%2074%25%20in%20January%202020].

Kemp, S. (2020c) 'Digital 2020: Uganda.' *Datareportal*, 18 February [https://datareportal.com/reports/digital-2020-uganda].

Kemp, S. (2020d) 'Digital 2020: Denmark.' *Datareportal*, 17 February [https://datareportal.com/reports/digital-2020-denmark].

Kendig, H. and McCallum, J. (1986) *Greying Australia: Future Impacts of Population Ageing*. Canberra: Australian Government Publishing Service.

Ketchell, M. (2018) 'The digital divide is not going away.' *The Conversation*, March 29 [https://theconversation.com/australias-digital-divide-is-not-going-away-91834].

Khoo, S-E., Birrell, B. and Heard, G. (2009) 'Intermarriage by birthplace and ancestry in Australia.' *People and Place* 17(1): 15–28.

Kimmel, M. (2016) *The Gendered Society*, 6th edition. Oxford: Oxford University Press.

Kimmel, M. (2004) 'A black woman took my job.' *New Internationalist*, 2 November [https://newint.org/features/2004/11/01/men].

King, N. and Calasanti, T. (2013) 'Men's aging amidst intersecting relations of inequality.' *Sociology Compass* 7(9): 699–710.

Kisilevich, S., Ang, C. and Last, M. (2012) 'Large-scale analysis of self-disclosure patterns among online social network users: A Russian context.' *Knowledge and Information Systems* 32(3): 609–628.

Kittay, E. F. (2013) *Love's Labor: Essays on Women, Equality and Dependency*. London and New York: Routledge.

Klinth, R. and Johannson, T. (2008) 'Caring fathers: The ideology of gender equality and masculine positions.' *Men and Masculinities* 11(1), October: 42–62.

Knight, K. (2013) 'No sex please—the new trend in online dating that specializes in platonic connections.' *Daily Mail*: Australia, 2 September [https://www.dailymail.co.uk/femail/article-2408482/Online-dating-platonicpartners-uk-creativesdating-com-lead-new-trend-platonic-connections.html].

Kocamaner, H. (2018) 'The politics of family values in Erdogan's new Turkey.' *Middle East Report* 288, Fall [https://merip.org/2018/12/the-politics-of-family-values-in-erdogans-new-turkey/].

Koleth, E. (2010) *Multiculturalism: A review of Australian policy statements and recent debates in Australia and overseas*. Research Paper no. 6 2010–11, 8 October. Canberra, Australia: Parliament of Australia [https://www.aph.gov.au/About_Parliament/Parliamentary_Departments/Parliamentary_Library/pubs/rp/rp1011/11rp06].

König, R., Seifert, A. and Doh, M. (2018) 'Internet use among older Europeans: An analysis based on SHARE data.' *Universal Access in the Information Society* 17: 621–633.

Kreager, D., Cavanagh, S., Yen, J. and Yu, M. (2014) '"Where have all the good men gone": Gendered interactions in online dating.' *Journal of Marriage and the Family* 76(2): 387–410.

Krekula, C., Arvidson, M., Heikkinen, S., Henriksson, A. and Olsson, E. (2017) 'On gray dancing: Constructions of age-normality through choreography and temporal codes.' *Journal of Aging Studies* 42: 38–45.

Krizanovich, K. (2018) 'Dating app Cougar . . . explains why she's using online dating apps to seek out hunky younger men.' *The Sun*, 21 June [https://www.thesun.co.uk/fabulous/5732709/im-53-divorced-and-swiped-right-on-your-son-says-dating-app-cougar/].

Kukunda, L. (2016) 'Why Ugandan men suck!' *Dispatch*, November 25 [https://www.dispatch.ug/2016/11/25/why-ugandan-men-suck/].

Lamidi, E., Brown, S. L. and Manning, W. D. (2015) 'Assortative mating: Racial homogamy in U.S. marriages, 1964–2014.' Family Profiles, FP-15–16. Bowling Green, OH: National Center for Family & Marriage Research [https://www.bgsu.edu/ncfmr/resources/data/family-profiles/lamidi-brown-manning-assortative-mat ing-racial-homogamy-fp-15-16.html].

Lamont, T. (2016) 'Life after the Ashley Madison affair.' *Guardian*, 28 February [https://www.theguardian.com/technology/2016/feb/28/what-happened-after-ashley-madi son-was-hacked].

Lampard, R. (2020) 'Meeting online or offline? Patterns and trends for co-resident couples in early 21st-century Britain.' *Sociological Review Online*, [https://doi.org/10.1 177%2F1360780419895524].

Lapidos, J. (2010) 'Profane friendship' *Slate*, October [http://www.slate.com/articles/life/strictly_platonic/2010/09/profane_friendship.html].

Larmour, C. (1999) 'Sex discrimination legislation in the States and Territories'. Research Paper 17 1998–1999, 30 March. Canberra: Parliament of Australia.

Laruni, E. (2015) 'Regional and ethnic identities: The Acholi of Northern Uganda, 1950–1968.' *Journal of East African Studies* 9(2): 212–230.

Lass, I. (2019) 'Language of love: A quarter of Australians are in inter-ethnic relationships.' *The Conversation*, July 30 [https://theconversation.com/language-of-love-a-quarter-of-australians-are-in-inter-ethnic-relationships-120416].

Laumann, E., Das, A. and Waite, L. (2008) 'Sexual dysfunction among older adults: Prevalence and risk factors from a nationally representative U.S. probability sample of men and women 57–85 years of age.' *The Journal of Sexual Medicine* 5(10): 2300–2311.

Laumann, E., Nicolosi, A., Glasser, D., Paik, A., Gingell, C., Moreira, E. and Wang, T. (2005) 'Sexual problems among women and men aged 40–80 years: Prevalence and correlates identified in the global study of sexual attitudes and behaviours.' *International Journal of Impotence Research* 17: 39–57.

Lawton, Z. and Callister, P. (2010) 'Older women–younger men relationships: The social phenomenon of "Cougars". A research note.' Victoria University of Wellington, New Zealand: Institute of Policy Studies Working Paper 10/02, January [https://www.researchgate.net/publication/260320175_Older_Women-Younger_Men_Relationships_the_Social_Phenomenon_of_'Cougars'_A_Research_Note].

Lee, J. (2019) 'Mediated superficiality and misogyny through cool on Tinder.' *Social Media+Society* 5(3), published online September 5 [https://journals.sagepub.com/doi/10.1177/2056305119872949].

Lea, M. and Spears, R. (1995) 'Love at first byte? Building personal relationships over computer networks', pp. 197–233. In J.T. Wood and S.W. Duck (eds), *Understanding Relationships: Off the Beaten Track.* Newbury Park, CA: Sage.

Leabeater, E. (2020) 'Supermarket trysts and video chats: How Australians are dating during Covid-19.' *The Guardian*, Friday 17, April [https://www.theguardian.com/lifeandstyle/2020/apr/17/supermarket-trysts-and-video-chats-how-australians-are-dating-during-covid-19].

Lee, S. and Niederle, M. (2011) 'Propose with a rose? Signaling in internet dating markets.' NBER Working Paper Series No. 17340. Cambridge MA: National Bureau of Economic Research.

Leggatt, J. (2017) 'Yes, it matters that you're sexually attracted to your partner.' *Sydney Morning Herald*, July 4 [https://www.smh.com.au/lifestyle/life-and-relationships/yes-it-matters-that-youre-sexually-attracted-to-your-partner-20170703-gx3kbh.html].

Lenhart, A. and Fox, S. (2006) 'Bloggers.' Report, Pew Research Centre: Internet & Technology [https://www.pewinternet.org/2006/07/19/bloggers/].

Lenton, A. and Francesconi, M. (2010) 'How humans cognitively manage an abundance of mate options.' *Psychological Science* 21(4): 528–533.

Levaro, E. B. (2011) *Theorizing Age and Gender in the Pursuit of Love in Late Life.* PhD dissertation, Department of Human Development and Family Studies. Eugene, Oregon, USA: University of Oregon.

Lewis, H. (2020) 'What it's like to be a Leftover woman.' *The Atlantic*, March 12 [https://www.theatlantic.com/international/archive/2020/03/leftover-women-china-israel-children-marriage/607768/].

Li, J. and Lipscomb, A. (2017) 'Love on the Cloud: The rise of online dating in China.' *US-China Today*, July 17 [https://uschinatoday.org/features/2017/07/17/love-on-the-cloud-the-rise-of-online-dating-in-china/].

Li, J., Zhu, A., Liu, D., Zhao, W., Zhou, Y., Chen, Y., Liu, Y. and Sun, N. (2020) 'Sustainability of China's singles day shopping festivals.' *Sustainability* 12(7) [https://www.mdpi.com/2071-1050/12/7/2644/htm].

Livingstone, G. and Brown, A. (2017) 'Intermarriage in the U.S. 50 years after Loving v. Virginia', May 18. Pew Research Centre [https://www.pewsocialtrends.org/2017/05/18/intermarriage-in-the-u-s-50-years-after-loving-v-virginia/].

Lodge, A. C. and Umberson, D. (2013) 'Age and embodied masculinities: Midlife gay and heterosexual men talk about their bodies.' *Journal of Aging Studies* 27: 225–232.

Lodge, A. C. and Umberson, D. (2012) 'All shook up: Sexuality of mid- to later life married couples.' *Journal of Marriage and Family* 74: 428–443.

Loe, M. (2004) 'Sex and the senior woman: Pleasure and danger in the Viagra era.' *Sexualities* 7(3): 303–326.

Lœgaard S. (2013) 'Danish anti-multiculturalism? The significance of the political framing of diversity', pp. 170–196. In P. Kivisto and Ö. Wahlbeck (eds), *Debating Multiculturalism in the Nordic Welfare States*, Palgrave Politics of Identity and Citizenship Series. London: Palgrave Macmillan.

Longmore, M. (1998) 'Symbolic interactionism and the study of sexuality.' *The Journal of Sex Research* 35(1): 44–57.

Lopreite, D. (2013) 'Gender policies in Argentina after neoliberalism: Opportunities and obstacles for women's rights.' *Latin American Perspectives* 42(1): 64–73.

Lowen, L. (2020) 'What does it mean when a woman is called a Cougar?' liveaboutdotcom, January 14 [https://www.liveabout.com/what-is-a-cougar-3534236].

Lubaale, G. (2019) 'Poverty in Uganda: Causes and strategies for reduction with great emphasis on ethics and ecological justice.' *Sociology and Anthropology* 7(1): 9–19.

Luo, S. (2017) 'Assortative mating and couple similarity: Patterns, mechanisms, and consequences.' *Social and Personality Psychology Compass* 11(8), DOI: 10.1111/spc3.12337.L.

Lupton, D. (2015) *Digital Sociology*. London and New York: Routledge.

Lutz, P. and Ranzini, G. (2017) 'Where dating meets data: Investigating social and institutional privacy concerns on Tinder.' *Social Media & Society* 3(1): 1–12.

Lykens, J., Pilloton, M., Silva, C., Schlamm, E., Wilburn, K. and Pence, E. (2019) 'Google for sexual relationships: Mixed-methods study on digital flirting and online dating among adolescent youth and young adults.' *JMIR Public Health Surveillance* 5(2), June 21 [https://pubmed.ncbi.nlm.nih.gov/31099335/].

McElfresh, A. (2020) 'Lasting love connections: How "just for fun" online dating profiles turned into marriages for two local couples.' *The Advocate*, March 8 [https://www.theadvocate.com/sponsored/baton_rouge/cox/article_3b699e58-5406-11ea-9cad-cbe2b1861a5b.html].

MacInnes, J. and Pérez Díaz, J. (2009) 'The reproductive revolution.' *The Sociological Review* 57(2): 262–284.

McIntosh, W. D, Locker, L., Briley, K., Ryan, R. and Scott, A. (2011) 'What do older adults seek in their potential romantic partners?: Evidence from online personal ads.' *International Journal of Aging and Human Development* 72(1): 67–82.

Mackenzie, C. and N. Stoljar (eds) (2000) *Relational Autonomy: Feminist Perspectives on Autonomy, Agency, and the Social Self.* Oxford: Oxford University Press.

McKenzie, L. (2015) *Age-Dissimilar Couples and Romantic Relationships: Ageless Love?* Basingstoke: Palgrave Macmillan.

McKeone, D. H. (1995) *Measuring your Media-Profile: A General Introduction to Media Analysis and PR Evaluation for the Communications Industry*. Hampshire, UK: Gower Press.

MacLeod, C. and McArthur, V. (2019) 'The construction of gender in dating apps: An interface analysis of Tinder and Bumble.' *Feminist Media Studies* 19(6): 822–840.

McMaster, D. (2001) *Asylum Seekers: Australia's Response to Refugees*. Melbourne: Melbourne University Press.

Macrotrends (2020) 'Kampala, Uganda Metro Area' [https://www.macrotrends.net/cities/22744/kampala/population#:~:text=The%20current%20metro%20area%20population,a%205.1%25%20increase%20from%202019].

McWilliams, S. and Barrett. A. E. (2012) 'Online dating in middle and later life: Gendered expectations and experiences.' *Journal of Family Issues* 35(3): 411–436.

Madson, L. M. (2013) '"High" and "low" in urban Danish speech styles.' *Language in Society* 42: 115–138.

Mafabi, D. and Oketch, M. L. (2017) 'Poor getting poorer, rich getting richer—report.' *Daily Monitor*, Thursday March 30 [https://www.monitor.co.ug/News/National/Poor—Uganda-rich—report/688334–3870102-jrr7ry/index.html].

MakingItMagazine (2013) 'Denmark: Stable, prosperous and happy.' [https://www.makingitmagazine.net/?p=7806].

Maldonado-Salcedo, M. (2017) 'Beauty weighs in Argentina: When looks can kill.' *Revista: Harvard Review of Latin America* [https://revista.drclas.harvard.edu/book/beauty-weighs-argentina].

Mallenbaum, C. (2016) 'Tinder update introduces 37 new gender identity options.' *The Sydney Morning Herald*, November 17 [https://www.smh.com.au/lifestyle/life-and-relationships/tinder-update-has-introduced-new-gender-identity-options-20161117-gsr710.html].

Malta, S. (2013) 'Why go online? Older adults' reasons for online dating.' *Reflections, Intersections and Aspirations, 50 years of Australian Sociology*, The Australian Sociological Association (TASA), Refereed Conference Proceedings [https://www.tasa.org.au/wp-content/uploads/2013/11/Malta.pdf].

Malta, S. (2007) 'Love Actually! Older adults and their romantic internet relationships.' *Australian Journal of Emerging Technologies & Society* 5(2): 84–102.

Malta, S. and Farquharson, K. (2014a) 'The initiation and progression of late-life romantic relationships.' *Journal of Sociology* 50(3): 237–251.

Malta, S. and Farquharson, K. (2014b) 'Online dating could have been made for older adults—they love it.' *The Conversation*, February 25 [https://theconversation.com/online-dating-could-have-been-made-for-older-adults-they-love-it-23344].

Malta, S., Temple-Smith, M., Bickerstaffe, A., Bourchier, L. and Hocking, J. (2020) ' "That might be a bit sexy for someone your age": Older adult sexual health conversations in primary care.' *Australasian Journal on Ageing* 39 (Supplement 1): 40–48.

Mansour, G. (2017) 'Looking for companionship and friendship.' SeniorsOnline Victoria, blog, 5 April [https://www.seniorsonline.vic.gov.au/News-Opinions/Blogs/Blog-Collection-Page/Looking-for-companionship-and-friendship].

Mantovani, G. (1994) 'Is computer-mediated communication intrinsically apt to enhance democracy in organizations?' *Human Relati*ons 47(1): 45–62.

Margaret (2020) 'We have different values! Can it work?' *At Table for Six (AT46) Newsletter*, January [https://atableforsix.com.au/we-have-different-values-can-it-work/].

Markus, A. (2019) *Mapping Social Cohesion*, 12th Report. Scanlon Foundation Research Institute [https://www.monash.edu/__data/assets/pdf_file/0004/2040268/mapping-social-cohesion-national-report-2019.pdf].

Marshall, B. (2018) 'Happily ever after? "Successful ageing" and the heterosexual imaginary.' *European Journal of Cultural Studies* 21(3): 363–381.

Marshall, B. (2010) 'Science, medicine and virility surveillance: "Sexy seniors" in the pharmaceutical imagination.' *Sociology of Health & Illness* 32(2): 211–224.

Marshall, B. (2008) 'Older men and sexual health: Post-Viagra views of changes in function.' *Generations* 32(1): 21–27.

Marshall, B. and Katz, S. (2002) 'Forever functional: Sexual fitness and the ageing male.' *Body & Society* 8(4): 43–70.

Marti, S. (2018) ' "The symbol of our nation": The slouch hat, the First World War, and Australian identity.' *Journal of Australian Studies* 42(1): 3–18.

Mason, B. (2019). 'Scott Morrison says multicultural Australia is like a "fragrant garam masala": PM explains multiculturalism'. *SBS News*, 22 October [https://www.sbs.com.au/news/scott-morrison-says-multicultural-australia-is-like-a-fragrant-garam-masala].

Matai, M-C. and Abrudan, M-M. (2017) 'Are national cultures changing? Evidence from the World Values Survey.' *Procedia—Social and Behavioral Sciences* 238: 657–664.

Mathews, M. (2019) 'Dating and menopause.' MegsMenopause (MM), blog, 15 May [https://megsmenopause.com/2019/05/15/dating-and-menopause/].

Matveyeva, A. (2020) 'Finding romance in gender-equal Denmark.' *P. S. I Love You*, May 10 [https://psiloveyou.xyz/finding-romance-in-gender-equal-denmark-7a0c9a9383fd].
Mayhew, S. (2015) 'queer geographies', pp. 409–410. *Oxford Dictionary of Geography*, 5th edition. Oxford: Oxford University Press.
Meah, A., Hockey, J. and Robinson, V. (2011) ' "I'm a sex kitten, aren't I . . .": Relocating agency and pleasure in older women's narratives about sex.' *Australian Feminist Studies* 26(67): 57–71.
Mearns, G. (2020) 'Queer geographies of spatial media.' *Geography Compass* 14(3): DOI: 10.1111/gec3.12481.
Megarry, J. (2014) 'Online incivility or sexual harassment? Conceptualising women's experiences in the digital age.' *Women's Studies International Forum* 47: 46–55.
Mellish, K. X. (2015) 'Dr. DK: "I er Danish passive med kærligheden".' *How to Live in Denmark: A Guide for Foreigners and their Danish Friends*, August 11 [https://www.howtoliveindenmark.com/category/dating-in-denmark/].
Mellström, U. (2009) 'The intersection of gender, race and cultural boundaries, or why is computer science in Malaysia dominated by women.' *Social Studies of Science* 39(6): 885–907.
Mellström, U. (2004) 'Machines and masculine subjectivity: Technology as an integral part of men's life experiences.' *Men and Masculinities* 6(4): 368–383.
Menard, D. A., Kleinplatz, P. J., Rosen, L., Lawless, S., Paradis, N., Campbell, M. and Huber, J. D. (2015) 'Individual and relational contributors to optimal sexual experiences in older men and women.' *Sexual and Relationship Therapy* 30: 78–93.
Menkin, J., Robles, T., Wiley, J. and Gonzaga, G. (2015) 'Online dating across the life span: Users' relationship goals.' *Psychology and Aging* 30(4): 987–993.
Mercks, A. (2015) YouTube: 'Muzungu—White privilege', September 29 [https://www.youtube.com/watch?v=Upw3T24qYog].
Mershon, A. M. (2019) 'Online dating after 60: Here are 4 ways to simplify your experience.' Sixty&Me: Senior Dating Sites [https://sixtyandme.com/tag/senior-dating-sites/].
Miguel, C. (2018) *Personal Relationships and Intimacy in the Age of Social Media*. Cham, Switzerland: Palgrave Pivot.
Miller, D., Costa, E., Haynes, N., McDonald, T., Nicolescu, R., Sinanan, J., Spyer, J., Venkatraman, S. and Wang, X. (2016) *How the World changed Social Media*. London: UCL Press.
Miller, E. L. (2019) 'Mind, body and boundaries: Self-presentation on the Nordic LGBTQ online dating scene.' *Frontiers in Psychology: Gender, Sex and Sexualities* 10 [https://doi.org/10.3389/fpsyg.2019.02770].
Milligan, C. and Passey, D. (2011) 'Ageing and the use of the internet: Current engagement and future needs.' State of the Art Review, Nominet Trust [http://nominettrust.org.uk/sites/default/files/NT%20SoA%20-%20Ageing%20and%20the%20use%20of%20the%20internet.pdf].
Mingle2 Uganda (no date) [https://mingle2.com/countries/346/Uganda].
Minichiello, V. and Coulsen, I. (2005) *Contemporary Issues in Gerontology: Promoting Positive Ageing*. Sydney: Allen & Unwin.
Minichiello, V., Browne, J. and Kendig, H. (2000) 'Perceptions and consequences of ageism: Views of older people.' *Ageing & Society* 20(3): 253–278.
Mintz, L. (2015) 'The orgasm gap: Simple truth and sexual solutions.' *Psychology Today*, October 4 [https://www.psychologytoday.com/au/blog/stress-and-sex/201510/the-orgasm-gap-simple-truth-sexual-solutions].

Mohdin, A. (2018) 'Meghan Markle's mixed marriage isn't unusual in the UK.' *Quartz*, May 17 [https://qz.com/1279306/royal-wedding-2018-meghan-markles-mixed-race-marriage-isnt-unusual-in-the-uk/].

Mølbak, R. (2010) 'Flirting as a liminal experience: A dialogue between a philosophical phenomenological understanding of liminality and an empirical–existential study of flirting.' Unpublished PhD Dissertation, McAnulty College and Graduate School of Liberal Arts, Duquesne University, USA.

Moore, A. (2019) 'How Tinder took me from serial monogamy to casual sex.' *The Guardian*, Sunday 28 September [https://www.theguardian.com/lifeandstyle/2014/sep/28/tinder-serial-monogamy-casual-sex].

Moore, N. (2012) 'The politics and ethics of naming: Questioning anonymisation in (archival) research.' *International Journal of Social Research Methodology* 15(4): 331–340.

Moore, N. (2007) '(Re) using qualitative data?' *Sociological Research Online* 12(3): 1–13.

Moran, A. (2011) 'Multiculturalism as nation-building in Australia: Inclusive national identity and the embrace of diversity.' *Ethnic and Racial Studies* 34(12), 25 May: 2153–2172.

Moreira, E. D., Glasser, D. B., King, R., Duarte, F. G. and Gingell, C. (2008) 'Sexual difficulties and help-seeking among mature adults in Australia: Results from the Global Study of Sexual Attitudes and Behaviours. *Sexual Health* 5: 227–234.

Moss, E. H. (2013) 'Online dating sucks for men because of women like me.' *Mamamia*, 4 March [http://www.mamamia.com.au/relationships/online-dating/].

Moss, R. (2020) 'Does Britain still have an issue with interracial relationships.' *The Huffington Post*, 18 January [https://www.huffingtonpost.co.uk/entry/life-in-mixed-race-relationships-britain_uk_5e173393c5b600960c612ddf].

Muise, A. (2011) 'Women's sex blogs: Challenging dominant discourses of heterosexual desire.' *Feminism & Psychology* 21(3): 411–419.

Murray, M. (2018) 'Older men dating younger women: What online dating data tells us.' *The Datemix*, February 21 [https://www.zoosk.com/date-mix/dating-advice/older-men-dating-younger-women/].

Mutluer, N. (2019) 'The intersectionality of gender, sexuality, and religion: Novelties and continuities in Turkey during the AKP era.' *Southeast European and Black Sea Studies* 19(1): 99–118.

muzungubloguganda (2014) 'How to date a Ugandan.' Diary of a Muzungu, August 6 [https://www.muzungubloguganda.com/society-and-culture/how-to-date-ugandan/].

Nannyonga-Tamusuza, S. (2009) 'Female-men, male-women, and others: Constructing and negotiating gender among the Baganda of Uganda.' *Journal of Eastern African Studies 3*(2): 367–380.

Nannyonga-Tamusuza, S. (2005) *Baakisimba: Gender in the Music and Dance of the Baganda People of Uganda*. New York and London: Routledge.

National Institute on Aging, US Dept of Health and Human Services, Health and Ageing (2017) 'HIV, AIDS, and older people', November 28 [http://www.nia.nih.gov/health/publication/hiv-aids-and-older-people].

Navarro, E. (2015) 'As the Ashley Madison fallout continues, are its members being misunderstood?' *CPH Post Online*, November 27 [cphpost.dk].

NBS (2016) 'Online dating on the rise in Uganda' [https://nbs.ug/2016/10/online-dating-on-the-rise-in-uganda/].

Neeland, J. (2019) 'Older Australians are embracing the world of online dating.' *HelloCare* [https://hellocaremail.com.au/older-australians-embrace-world-online-dating/].

Neumann, K. and Tavan, G. (eds) (2013) *Does History Matter: Making and Debating Citizenship, Immigration and Refugee Policy in Australia and New Zealand.* Canberra, Australia: ANU Press.

Ng, D. and Ning, T. J. (2018) 'These are the "leftover men" of China, who just want to get married.' *CNA Insider* (Channel News Asia), 30 June [https://www.channelnewsasia.com/news/cnainsider/leftover-men-china-get-married-gender-imbalance-one-child-policy-10485358].

Nguyen, G. A-N. (2018) 'Why I'm glad I swiped right on so many "failed" dates.' *SBS*, 2 August [https://www.sbs.com.au/topics/voices/relationships/article/2018/07/31/why-im-glad-i-swiped-right-so-many-failed-dates].

Noone, Y. (2012) 'Safe sex campaign for seniors.' [http://www.australianageingagenda.com.au/2012/09/27/safe-sex-campaign-for-seniors/].

NordeaTrade (2020) 'The economic context of Denmark' [https://www.nordeatrade.com/dk/explore-new-market/denmark/economical-context].

Novak, L. (2014) 'Pre-teen boys worried about body image and muscles, Flinders University study finds.' *The Advertiser*, December 28 [https://www.adelaidenow.com.au/news/south-australia/preteen-boys-worried-about-body-image-and-muscles-flinders-university-study-finds/news-story/80e24eacf63e33c1dc4fc1fe60fe3d44].

Office for National Statistics (2019a) 'Internet users, UK: 2019: Internet use in the UK annual estimates by age, sex, disability and geographical location.' *Statistical Bulletin*, 24 May. [https://www.ons.gov.uk/businessindustryandtrade/itandinternetindustry/bulletins/internetusers/2019#generation-gap-narrowing-in-recent-internet-use].

Office for National Statistics (2019b) 'National life tables, UK: 2016 to 2018.' [https://www.ons.gov.uk/peoplepopulationandcommunity/birthsdeathsandmarriages/lifeexpectancies/bulletins/nationallifetablesunitedkingdom/2016to2018].

Officer, A., Schneiders, M., Wu, D., Nash, P., Thiyagarajan, J. A. and Beard, J. R. (2016) 'Valuing older people: Time for a global campaign to combat ageism.' *Bulletin of the World Health Organisation* 94: 710–710A.

OHCHR (Office of the United Nations High Commissioner for Human Rights) (no date) 'Child, early and forced marriage, including in humanitarian settings.' [https://www.ohchr.org/EN/Issues/Women/WRGS/Pages/ChildMarriage.aspx].

OkCupid (2016) 'A digital decade: Sex.' *The Deep End* [https://www.okcupid.com/deep-end/digital-decade-sex].

Okeleke, K. (2018) *Uganda: Driving Inclusive Socio-economic Progress through Mobile-enabled Digital Transformation.* London: GSMA: Global System for Mobile Communications.

Ong, D. and Wang, J. (2015) 'Income attraction: An online dating field experiment.' *Journal of Economic Behaviour & Organisation* 111: 13–22.

Open Eye (2019) 'Do Victoria Milan use fake female users to engage male users?' *Medium*, August 24 [https://medium.com/@openeyes.journalism/the-interesting-story-of-victoria-milan-5f1cf53bfbaf].

Orange, R. and Duncan, P. (2019) 'And the least feminist nation in the world is . . . Denmark?' *The Guardian*, Friday 10 May [https://www.theguardian.com/lifeandstyle/2019/may/10/and-the-least-feminist-nation-in-the-world-is-denmark].

Ortega, J. and Hergovich, P. (2017) 'The strength of absent ties: Social integration via online dating', arXiv:1709.10478v1[physics.soc-ph].

Østergård, U. (2012) 'Danish national identity', pp. 37–55. In M. C. Gertson, A. M. Søderberg and M. Zølner (eds), *Global Collaboration: Intercultural Experiences and Learning*. London: Palgrave Macmillan.

Østergård, U. (2000) 'Danish national identity: Between multinational heritage and small state nationalism', pp. 139–184. In B. Branner and M. Kelstrup (eds), *Denmark's Policy towards Europe after 1945*. Odense, Denmark: Odense University Press.

Oswaks, M. (2015) '4 couples share their online dating success stories', *Cosmopolitan*, March 2 [https://www.cosmopolitan.com/sex-love/a36793/couples-share-their-online-dating-success-stories/].

Owens, J. and Sirois, F. (2019) 'Review of the impact of loneliness and social isolation on health and well-being and whether people who experience loneliness/social isolation have higher use of public services'. Cardiff: Welsh Government, GSR report number 46/2019 [https://gov.wales/loneliness-and-use-public-services-literature-review].

Oxford English Dictionary Online (no date) Definition of cisgender [http://oxforddictionaries.com/definition/english/cisgender].

Paasonen, S., Light, B. and Jarrett, K. (2019) 'The dick pic: Harassment, curation, and desire.' *Social Media + Society* 5(2), DOI: 10.1177/2056305119826126.

PAI: Executive Summary, Uganda (2010) 'The effects of a very young age structure in Uganda.' Population Action International, The Shape of Things to Come series [https://web.archive.org/web/20130124120235/http://www.populationaction.org/oldmedia/SOTC_Uganda.pdf].

Paisley, E. (2018) 'What to know about the online dating landscape in 2018.' globalwebindex, October 16 [https://blog.globalwebindex.com/trends/online-dating/].

Parker, C. P. and Carden, K. (2018) 'Online dating in the golden years.' *Psychology Today Australia*, November 3 [https://www.psychologytoday.com/au/blog/engaging/201811/online-dating-in-the-golden-years].

Pasca, T. (2018) 'Online dating apps have a major problem with sexual harassment—but solutions must start offline.' *This Magazine*, September–October, September 27 [https://this.org/2018/09/27/online-dating-apps-have-a-major-problem-with-sexual-harassment-but-solutions-must-start-offline/].

Pascoe, C. J. (2011) 'Resource and risk: Youth sexuality and new media use.' *Sexuality Research and Social Policy* 8(1): 5–17.

Patel, A. (2018) 'Dating multiple people at once is the norm—here's how to do it right.' *Global News*, January 4 [https://globalnews.ca/news/3943295/dating-more-than-one-person/].

PerCapita: Fighting Inequality in Australia (2020) 'Measure for measure: Gender equality in Australia', 3 March [https://percapita.org.au/our_work/measure-for-measure-gender-equality-in-australia/].

Perkins, S. (2015) *Hegemonic Masculinity and its Effects on Attitudes towards Seeking Professional Psychological Help*. PhD dissertation, Counselling Psychology. Boston, Ma.: Northeastern University.

Petter, O. (2018) 'Racism is rife on dating apps—Where does it come from and how can it be fixed?' *The Independent*, Friday 24 August [https://www.independent.co.uk/life-style/love-sex/dating-apps-racism-tinder-bumble-grindr-online-dating-a8504996.html].

Pew Research Center (2020) 'Men who have online dated in the past five years are more likely than women to say they didn't get enough messages', February 6 [https://www.pewresearch.org/ft_2020-02-06_datingtakeaways_05/].

Pew Research Center (2019) 'Household patterns by age and gender. Religion and public life' [https://www.pewforum.org/2019/12/12/household-patterns-by-age-and-gender/].
Philbeck, J. (1997) 'Seniors and the internet.' *Cybersociology*, 2, August [https://www.cybersociology.com/files/2_2_philbeck.html].
Phillips, J. (2015) 'Asylum seekers and refugees: What are the facts?' Canberra, Australia: Parliament of Australia [https://www.aph.gov.au/About_Parliament/Parliamentary_Departments/Parliamentary_Library/pubs/rp/rp1415/AsylumFacts].
Phoenix, A. (2013) 'Analysing narrative context', pp 72–87. In Andrews, M. Squire, C. and Tamboukou, M. (eds), *Doing Narrative Research*. Los Angeles, CA: Sage.
Pidd, H. (2015) 'Sexual activity survey debunks myths concerning lives of older people.' *The Guardian*, 29 January [https://www.theguardian.com/uk-news/2015/jan/28/older-sexual-activity-ageing-survey-manchester].
Pinsker, J. (2020) 'When does someone become "Old": It's surprisingly hard to find a good term for people in late life.' *The Atlantic*, January 27 [https://www.theatlantic.com/family/archive/2020/01/old-people-older-elderly-middle-age/605590/].
Planned Parenthood Federation of America (PPFA) (2007) *The Health Benefits of Sexual Expression White Paper* [https://www.plannedparenthood.org/files/3413/9611/7801/Benefits_Sex_07_07.pdf 13].
Planned Parenthood Federation of America (2003). *White Paper: The Health Benefits of Sexual Expression* [http://www.plannedparenthood.org/files/PPFA/BenSex_07-07.pdf].
Platformglobig UK (2020) 'Population—Ethnicity, language in the UK', 31 January [https://platform.globig.co/knowledgebase/UK/landscape/population-ethnicity-language-in-theuk#:~:text=While%20the%20UK%20has%20a,cultures%2C%20it%20is%20overwhelmingly%20white.&text=The%20official%20language%20of%20the,who%20do%20not%20speak%20it].
Plecher, H. (2020) 'Urbanization in the United Kingdom 2019.' Statista, July 15 [https://www.statista.com/statistics/270369/urbanization-in-the-united-kingdom/].
PopulationStat (2020) 'Kampala, Uganda population. [https://populationstat.com/uganda/kampala].
Potârcă, G. (2014) 'Modern love: Comparative insights in online dating preferences and assortative mating.' Unpublished PhD thesis, University of Groningen, the Netherlands.
Potârcă, G. and Mills, M. (2015) 'Racial preferences in online dating across European countries.' *European Sociological Review* 31(3): 326–341.
Potter-Collins, A. (2014) '2011 Census analysis: What does the 2011 Census tell us about inter-ethnic relationships?' Office for National Statistics, UK [https://www.ons.gov.uk/peoplepopulationandcommunity/birthsdeathsandmarriages/marriagecohabitationandcivilpartnerships/articles/whatdoesthe2011censustellusaboutinterethnicrelationships/2014-07-03].
Pratt, G. and Rosner, V. (2006) 'Introduction: The global & the intimate.' *Women's Studies Quarterly* 34(1/2), Spring–Summer: 13–24.
Prensky, M. (2001) 'Digital natives, digital immigrants part 1.' *On the Horizon* 9(5): 1–6.
Prescott, J., Gray, N., Smith, F. and McDonagh, J. (2015) 'Blogging as a viable research methodology for young people with arthritis: A qualitative study.' *Journal of Medical Internet Research* 17(3): DOI: 10.2196/jmir.3608.
Prestage, G. P., Bavinton, B., Grierson, J., Down, I., Keen, P., Bradley, J. and Duncan, D. (2015) 'Online dating among Australian gay and bisexual men: Romance or hooking up?' *AIDS and Behaviour* 19(10), March: 1905–1913.

Price, J. (2015) 'Dating at our age—Is erectile dysfunction the problem?' *Senior Planet*, online community website, sponsored by Older Adults Technology Services (OATS), USA, Tuesday January 6 [https://seniorplanet.org/dating-at-our-age-is-erectile-dysfunction-the-problem/].

Prince, E. (no date) 'Telo sex hotels: The best kept secret in Buenos Aires.' *IExplore* [https://www.iexplore.com/destinations/buenos-aires/telo-hotels].

Qin, A. (2019) 'A prosperous China says "men preferred", and women lose.' *The New York Times*, July 16 [https://www.nytimes.com/2019/07/16/world/asia/china-women-discrimination.html].

Quattrociocchi, W. (2016) 'How does misinformation spread online?' World Economic Forum, 14 January [https://www.weforum.org/agenda/2016/01/q-a-walter-quattrociocchi-digital-wildfires/].

Rachel (2014) 'What does it mean to be Australian?' Rachel, blog, January 7 [https://rachel.blog/2014/01/07/what-does-it-mean-to-be-australian/].

Rae, A. (2018) 'Think your country is crowded? These maps reveal the truth about population density across Europe.' *The Conversation*, January 24 [https://theconversation.com/think-your-country-is-crowded-these-maps-reveal-the-truth-about-population-density-across-europe-90345].

Rainie, L. and Wellman, B. (2012) *Networked*. Cambridge, MA: The MIT Press.

Rajadurai, E. (2018) 'Why Australia is the world's most successful multicultural society.' The McKell Institute, December [https://mckellinstitute.org.au/research/articles/why-australia-is-the-worlds-most-successful-multicultural-society/].

Ramirez, A., Sumner, E. M., Fleuriet, C. and Cole, M. (2015) 'When online dating partners meet offline: The effect of modality switching on relational communication between online daters.' *Journal of Computer-Mediated Communication* 20(1): 99–114.

Rege, A. (2009) 'What's love got to do with it? Exploring online dating scams and identity fraud.' *International Journal of Cyber Criminology* 3(2): 494–512.

Regnerus, M. and Uecker, J. (2011) *Premarital Sex in America: How Young Americans Meet, Mate, and Think about Marriage*. New York and Oxford: Oxford University Press.

Reis, H. T. (2012) 'A brief history of relationship research in social psychology', pp. 363–382. In Kruglanski, A. W. and Stroebe, W. (eds), *Handbook of the History of Social Psychology*. New York: Psychology Press.

Relationships Australia (RASA) (2020) Relationship advice sheet: 'Online relationships' [https://www.relationships.org.au/relationship-advice/relationship-advice-sheets/online-relationships].

Relationships Australia (RASA) (2017) 'Online dating: Results.' November 2017, monthly online survey [https://www.relationships.org.au/what-we-do/research/online-survey/november-2017-online-dating/results].

Rhohlinger, D. A. (2002) 'Eroticizing men: Cultural influences on advertising and male objectification.' *Sex Roles* 46(3/4): 61–74.

Ricciardelli, R., Clow, K. A. and White, P. (2010) 'Investigating hegemonic masculinity: Portrayals of masculinity in men's lifestyle magazines.' *Sex Roles* 63(1–2): 64–78.

Rich, A. (1980) 'Compulsory heterosexuality and lesbian existence.' *Signs* 5(4): 631–660.

Richardson, D. (1996) 'Heterosexuality and social theory', pp. 1–21. In Richardson, D. ed. *Theorizing Heterosexuality*. Buckingham, England: Open University Press.

Robnett, B. and Feliciano, C. (2011) 'Patterns of racial–ethnic exclusion by internet daters.' *Social Forces* 89(3): 807–828.

Roscigno, V., Mong, S., Byron, R. and Tester, G. (2007) 'Age discrimination, social closure and employment.' *Social Forces* 86(1): 313–334.

Rosenbloom, S. (2011) 'Love, lies and what they learned.' *The New York Times*, November 12 [https://www.nytimes.com/2011/11/13/fashion/online-dating-as-scientific-research.html].

Roseneil, S. (2007) 'Living and loving beyond the heteronorm: A queer analysis of personal relationships in the twenty-first century.' *Eurozine*, 29 May [https://www.eurozine.com/living-and-loving-beyond-the-heteronorm/].

Roseneil, S. (2005) 'Living and loving beyond the boundaries of the heteronorm: Personal relationships in the 21st century', pp. 241–258. In L. McKie, Cunningham-Burley, S. and McKendrick, J. (eds), *Families in Society: Boundaries and Relationships*. Bristol: Policy Press.

Roseneil, S. (2004) 'Why we should care about friends: An argument for queering the care imaginary in social policy.' *Social Policy & Society* 3(4): 409–419.

Roseneil, S. and Budgeon, S. (2004) 'Cultures of intimacy and care beyond the "Family": Personal life and social change in the early 21st century.' *Current Sociology* 52(2): 135–159.

Rosenfeld, M. (2018) 'Are Tinder and dating apps changing dating and mating in the U.S.?', pp. 103–117. In J. Van Hook, S. McHale and V. King (eds), *Families and Technology*. Cham, Switzerland: Springer.

Rosenfeld, M. and Thomas, R. (2012) 'Searching for a mate: The rise of the Internet as a social intermediary.' *American Sociological Review* 77(4): 523–547.

Rosengren, K. E. (1981) 'Advances in Scandinavian content analysis: An introduction', pp. 9–19. In K. Rosengren (ed.), *Advances in Content Analysis*. Beverly Hills, CA: Sage.

Ross, M. W. (2005) 'Typing, doing, and being: Sexuality and the internet.' *Journal of Sex Research*, 42(4): 342–352.

Rowland, K. (2020) 'What I learned talking to 120 women about their sex lives and desires.' *The Guardian*, 5 February [https://www.theguardian.com/lifeandstyle/2020/feb/05/katherine-rowland-the-pleasure-gap-women-sex-lives-desire].

Rowntree, M. (2014) '"Comfortable in my own skin": A new form of sexual freedom for ageing baby boomers.' *Journal of Aging Studies* 31, December: 150–158.

RSVP Date of the Nation Report 2015 [http://www.medianet.com.au/releases/release-details/?id=838333].

RSVP team (2016) '2016: It's a brave new world for dating', 15 January. [https://www.datehub.com.au/].

Rubin, G. 1984. 'Thinking sex: Notes for a radical theory for the politics of sexuality', pp 267–319. In C. Vance (ed.), *Pleasure and Danger: Exploring Female Sexuality*. Boston, MA: Routledge.

Rudder, C. (2014a) 'Race and attraction, 2009–2014.' *Oktrends: Dating Research from OkCupid*, September 10 [http://blog.okcupid.com/index.php/race-attraction-2009-2014/].

Rudder, C. (2014b) *Dataclysm: Who We Are (when we think no one's looking)*. New York: Random House LLC.

Ruggiero, T. E. (2000) *Uses and Gratifications Theory in the 21st Century*. Princeton, New Jersey: Princeton University Press.

Saito, F. (2007) 'Modernization theory revisited: Does a grand theory of Inglehart and Welzel really bring human development?' Seminar powerpoint [https://home.hiroshima-u.ac.jp/~hicec/coe/coe/PPT/seminar34-1.pdf].

Sakkal, P. (2020) 'Bettina Arndt Australia Day honour to be reviewed.' *The Sydney Morning Herald*, February 24 [https://www.smh.com.au/national/bettina-arndt-australia-day-honour-to-be-reviewed-20200224-p543ox.html].

Salmon, (2020) 'Is online dating the best way to find a partner?' *Daily Monitor*, Thursday March 26 [https://www.monitor.co.ug/artsculture/Heart-to-Heart/Is-online-dating-best-way-to-find-a-partner/691230-5504556–23cef6z/index.html].

Sandberg, L. (2015) 'In lust we trust? Masculinity and sexual desire in later life.' *Men and Masculinities* 19: 192–208.

Sandberg, L. (2013) 'Just feeling a naked body close to you: Men, sexuality and intimacy in later life.' *Sexualities* 16(3/4): 261–282.

Sandberg, L. (2009) 'Getting intimate: Old age, masculinity and new (?) heterosexual morphologies', pp. 61–78. *Gexcel Work in Progress Report Volume V. Proceedings from GEXcel Theme 2: Deconstructing the Hegemony of Men and Masculinities*. Linköping, Sweden: Linköping University [http://www.genderexcel.org/webfm_send/51].

Sargent-Cox, K. (2017) 'Ageism: We are our own worst enemy.' *International Psychogeriatrics* 29(1): 1–8.

Sari, Ç. (2019) '#10YearChallenge sparks hijab debate.' *Inside Turkey*, 17 August [https://insideturkey.news/2019/04/25/10yearchallenge-sparks-hijab-debate-cs/].

Sassler, S. (2010) 'Partnering across the life course: Sex, relationships and mate selection.' *Journal of Marriage and Family* 72(3): 557–575.

Sassler, S. and Miller, A. J. (2011) 'Waiting to be asked: Gender, power, and relationship progression among cohabiting couples.' *Journal of Family Issues* 32(4): 482–506.

Sassler, S., Michelmore, K. and Qian, Z. (2018) 'Transitions from sexual relationships into cohabitation and beyond.' *Demography* 55(2): 511–534.

Savage, M. (2019) 'The Nordic way: The "paradox" of working in the world's most equal countries.' *BBC*, Worklife, 5 September [https://www.bbc.com/worklife/article/20190831-the-paradox-of-working-in-the-worlds-most-equal-countries].

Sawlani, S. (2012) 'Uganda: When white is better than black—A Mzungu.' *The Observer* (Kampala), 26 April, AllAfrica [https://allafrica.com/stories/201204270046.html].

SBS (2013a) 'A look at ethnic dating in Australia.' *SBS News*, 26 August [https://www.sbs.com.au/news/a-look-at-ethnic-dating-in-australia_1].

SBS (2013b) 'For some Australians, maintaining a sense of cultural identity remains important in their choice of partner.' *SBS World News Australia Radio*, 26 August [https://www.sbs.com.au/news/a-look-at-ethnic-dating-in-australia_1].

Scheuermann, L. and Taylor, G. (1997) 'Netiquette.' *Internet Research* 7(4): 269–273.

Schwartz, J. P., Grammas, D. L., Sutherland, R. J., Stiffert, K. J. and Bush-King, I. (2010) 'Masculine gender roles and differentiation: Predictors of body image and selfobjectification in men.' *Psychology of Men & Masculinity* 11(3): 208–224.

Scott, I. (2016) 'A brief history of cyberfeminism', October 13[https://www.artsy.net/article/artsy-editorial-how-the-cyberfeminists-worked-to-liberate-women-through-the-internet].

Scott, S., Dawson, M. and Newmahr, S. (2014) 'A symbolic interactionist approach to asexuality', pp. 215–226. In T. Weinberg and S. Newmahr (eds), *Selves, Symbols, and Sexualities: An Interactionist Anthology*. Los Angeles: Sage.

Seidman, G. (2019) 'Can using online dating apps be addictive?' *Psychology Today*, August 21 [https://www.psychologytoday.com/au/blog/close-encounters/201908/can-using-online-dating-apps-be-addictive].

Seidman, S. (2009) 'Critique of compulsory heterosexuality.' *Sexuality Research and Social Policy: Journal of the National Sexuality Resource Center* 6: 18–28.

Seidman, S. (1991) *Romantic Longings: Love in America, 1830–1980*. New York and London: Routledge.

Sela, Y., Pham, M. N., Mogilski, J. K., Lopes, G. S., Shackelford, T. K. and Zeigler-Hill, V. (2018) 'Why do people disparage May–December romances? Condemnation of age-discrepant romantic relationships as strategic moralization.' *Personality and Individual Differences* 130: 6–10.

Serferou, A. (2019) 'All the Scandinavian dating rules you should know.' *Culture Trip*, 6 May [https://theculturetrip.com/europe/denmark/articles/all-the-scandinavian-dating-rules-you-should-know/].

Seferou, A. (2017) 'Couples in this European country are most likely to cheat.' *Culture Trip*, 8 August [https://theculturetrip.com/europe/denmark/articles/couples-in-this-european-country-are-most-likely-to-cheat/].

Selmer, J. and Lauring, J. (2013) 'Globalizing Denmark: Open-minded while ethnocentric.' *International Studies of Management & Organization* 43(1): 9–25.

Ševčíková, A. and Sedláková, T. (2020) 'The role of sexual activity from the perspective of older adults: A qualitative study.' *Archives of Sexual Behaviour* 49: 969–981.

Seymour, W. and Lupton, D. (2004) 'Holding the line online: Exploring online relationships for people with disabilities.' *Disability & Society* 19(4): 291–305.

Shade, L. R. (2002) *Gender and Community in the Social Construction of the Internet*. New York: Peter Lang.

Shaevitz, M. H. (2015) 'New relationships and sex after 60, 70, and 80.' *Psychology Today*, July 22 [https://www.psychologytoday.com/au/blog/refire-don-t-retire/201507/new-relationships-and-sex-after-60-70-and-80].

Shahbaz, A. (2018) 'Freedom on the Net 2018: The rise of digital authoritarianism' [https://freedomhouse.org/report/freedom-net/2018/rise-digital-authoritarianism].

Shaw, D. (2020) 'Why mature dating apps are coming.' *BBC News*, 1 January [https://www.bbc.com/news/business-50407218].

Sheftalovich, Z. (2014) 'Can you really find true love online?' *Choice: Online dating site and app reviews*, Online dating case studies [https://www.choice.com.au/electronics-and-technology/internet/using-online-services/articles/online-dating-sites-review].

Sheftalovich, Z. and Smith, G. (2020a) 'Online dating site and app reviews.' *Choice* (Consumer Advocacy magazine), 11 August [https://www.choice.com.au/electronics-and-technology/internet/using-online-services/articles/online-dating-sites-review].

Sheftalovich, Z. and Smith, G. (2020b) 'Online dating site and app reviews.' *Choice*, 11 February [https://www.choice.com.au/electronics-and-technology/internet/using-online-services/articles/online-dating-sites-review].

Sheppard, J. and Biddle, N. (2015) 'Is Australia as egalitarian as we think it is?' *ABC News*, Wednesday 28 October [https://www.abc.net.au/news/2015-10-28/sheppard-is-australia-as-egalitarian-as-we-think-it-is/6889602].

Showden, C. R. (2012) 'Theorising maybe: A feminist/queer theory convergence.' *Feminist Theory* 13(1): 3–25.

Shpancer, N. (2014) 'Laws of attraction: How do we select a life partner?' *Psychology Today*, December 2 [https://www.psychologytoday.com/us/blog/insight-therapy/201412/laws-attraction-how-do-we-select-life-partner].

Shpancer, N. (2012) 'The Cougar conundrum.' *Psychology Today*, October 4 [https://www.psychologytoday.com/us/blog/insight-therapy/201210/the-cougar-conundrum].

Siibak, A. (2010) 'Constructing masculinity on a social networking site. The case-study of visual self-presentations of young men on the profile images of SNS Rate.' *Young* 18(4): 403–425.

Skeggs, B. (2005) 'The making of class and gender through visualizing moral subject formation.' *Sociology* 39 (5): 965–982.

Skopek, J., Schmitz, A. and Blossfeld, H-P. (2011) 'The gendered dynamics of age preferences—Empirical evidence from online dating', *Journal of Family Research (Zeitschrift fur Familienschung)* 23(3): 267–290.

Smaill, B. (2004) 'Online personals and narratives of the self: Australia's RSVP.' *Convergence*, 10(1): 93–107.

Smart, C. (2007) Personal Life. Cambridge: Polity.

Smart, C. and Shipman, B. (2004) 'Visions in monochrome: Families, marriage and the individualization thesis.' *The British Journal of Sociology* 55(4): 491–509.

Smith, A. and Anderson, M. (2016) '5 facts about online dating.' *FACTANK: News in the Numbers*, February 29 [https://www.pewresearch.org/fact-tank/2016/02/29/5-facts-about-online-dating/].

Smith, A. and Duggan, M. (2013) 'Online dating and relationships.' Pew Research Center's Internet & American Life Project [http://www.pewinternet.org/2013/10/21/online-dating-relationships/].

Smith, L. (2018) 'What is a quasiplatonic aka queerplatonic relationship?' *PinkNews*, April 18 [https://www.pinknews.co.uk/2018/04/18/what-is-a-quasiplatonic-aka-queerplatonic-relationship-friendship/].

Smith, R. (2013) 'Why Turkey lifted its ban on the Islamic headscarf.' *National Geographic*, October 12 [https://www.nationalgeographic.com/news/2013/10/131011-hijab-ban-turkey-islamic-headscarf-ataturk/].

Snee, H. (2013) 'Framing the other: Cosmopolitanism and the representation of difference in overseas gap year narratives.' *British Journal of Sociology* 64(1): 142–162.

Soergel, A. (2020) 'Minding the Nordic inequality gap.' *U.S.News: A World Report*, January 16 [https://www.usnews.com/news/best-countries/articles/2020-01-16/gender-equality-perceptions-versus-reality-in-nordic-countries].

Song, M. (2015) 'What constitutes intermarriage for multiracial people in Britain?' *The Annals of the American Academy of Political and Social Science* 662 (1): 94–111.

Sorooshyari, N. (2011) 'The tensions between feminism and libertarianism: A focus on prostitution.' *Washington University Jurisprudence Review* 3(1): 167–193.

Soutphommasane, T. (2016) 'The success of Australia's multiculturalism.' Speech by the Race Discrimination Commissioner, Dr. Tim Soutphommasane, to the Sydney Institute, 9 March [https://humanrights.gov.au/about/news/speeches/success-australias-multiculturalism].

Spijker, J. and MacInnes, J. (2013) 'Population ageing: The timebomb that isn't?' *British Medical Journal* 347 (7933): 20–22. DOI: https://doi.org/10.1136/bmj.f6598.

Sproull, L. and Kiesler, S. (1991) *Connections: New Ways of Working in the Networked Organization*. Cambridge, Mass.: MIT Press.

Sproull, L. and Kiesler, S. (1986) 'Reducing social context cues: Electronic mail in organizational communication.' *Management Science* 32: 1492–1512.

Squire, C., Andrews, M. and Tamboukou, M. (2013) *Doing Narrative Research*. London: Sage.

Squires, C. (2013) 'From experience-centred to socioculturally-oriented approaches to narrative', pp 47–71. In M. Andrews, C. Squire, and M. Tamboukou (eds), *Doing Narrative Research*. Los Angeles, California: Sage.

Stacey, J. (2004) 'Cruising to familyland: Gay hypergamy and rainbow kinship.' *Current Sociology* 52(2): 181–197.

Stafford, T. F., Stafford, M. R. and Schkade, L. L. (2004) 'Determining uses and gratifications for the internet.' *Decision Sciences* 35: 259–288.

Stanley, L. (2008) 'Madness to the method? Using a narrative methodology to analyse large-scale complex social phenomena.' *Qualitative Research* 8(3): 435–447.

Stanley, L. and Sereva, E. (2019) 'Documents of life'. In P. Atkinson, S. Delamont, A. Cernat, J. W. Sakshaug and R. A. Williams (eds), SAGE *Research Methods Foundations*. Doi: 10.4135/9781526421036752360.

Stanley, L. and Wise, S. (2006) 'Putting it into practice: Using feminist fractured foundationalism in researching children in the concentration camps of the South African war.' *Sociological Research Online* 11(1) [http://www.socresonline.org.uk/11/1/stanley.html].

Statista (2020a) 'Online dating.' [https://www.statista.com/outlook/372/online-dating].

Statista (2020b) 'Share of people who have cheated on their current partner in Denmark, 2018, by gender.' Statista Research Department, August 25 [https://www.statista.com/statistics/1092815/share-of-people-who-have-cheated-on-their-current-partner-in-denmark-by-gender/].

Statista (2020c) 'United Kingdom: Degree of urbanization from 2009 to 2019' [https://www.statista.com/statistics/270369/urbanization-in-the-united-kingdom/].

Stephens, J. (1976) *Loners, Losers, and Lovers: Elderly Tenants in a Slum Hotel*. Oxford, UK: University of Washington Press.

Stephure, R., Boon, S. D, MacKinnon, S. L. and Deveau, V. L. (2009) 'Internet initiated relationships: Association between age and involvement in online dating.' *Journal of Computer-Mediated Communication* 14(3): 658–681.

Stevens, L. and Ostberg, J. (2012) 'Gendered bodies: Representations of femininity and masculinity in advertising practices.' In L. M. Visconti, L. Peñaloza and N. Toulouse (eds), *Marketing Management: A Cultural Perspective*. London: Routledge.

Stokel-Walker, C. (2018) 'Why is it OK for online daters to block whole ethnic groups.' *The Guardian*, Sunday 30 September [https://www.theguardian.com/technology/2018/sep/29/wltm-colour-blind-dating-app-racial-discrimination-grindr-tinder-algorithm-racism].

Su, X. and Hu, H. (2019) 'Gender-specific preference in online dating.' *EPJ Data Science* 8(12) [https://doi.org/10.1140/epjds/s13688-019-0192-x].

Sugiura, L., Wiles, R. and Pope, C. (2016) 'Ethical challenges in online research: Public/private perceptions.' *Research Ethics* 13(3–4): 184–199.

Swinburne University (2016) 'Loneliness can lead to poor mental health.' Media Release 1 July.

Tannenbaum, C., Ellis R.P., Eyssel, F., Zou, J., and Schiebinger, L. (2019) 'Sex and gender analysis improves science and engineering', Nature 575: 137–146.

Tarrant, A. (2010) '"Maturing" a sub-discipline: The intersectional geographies of masculinities and old age.' *Geography Compass* 4(10): 1580–1591.

Tawat, M. (2018) 'Multiculturalism: Is Denmark a den of intolerance and Sweden a land of political correctness?' London School of Economics, blog, June 12 [https://blogs.lse.ac.uk/religionglobalsociety/2018/06/multiculturalism-is-denmark-a-den-of-intolerance-and-sweden-a-land-of-political-correctness/].

Taylor E. A. and Scott, J. (2018) 'Gender: New consensus or continuing battleground', pp. 56–85. In Phillips, D., Curtice, J., Phillips, M. and Perry, J. (eds), *British Social Attitudes: The 35th Report*. London: The National Centre for Social Research.

Tepanon, Y. (2006) 'Exploring the minds of sex tourists: The psychological motivation of liminal people.' Unpublished PhD dissertation, Hospitality and Tourism Management. Blacksberg, Virginia, USA: Faculty of the Virginia Polytechnic Institute and State University.

Terry, R. (2019) '10 facts about life expectancy in Denmark.' The Borgen Project, blog, February 9 [https://borgenproject.org/10-facts-about-life-expectancy-in-denmark/].

Textor, C. (2020) 'Population in China 2008–2018, by gender.' *Statista*, May 26 [https://www.statista.com/statistics/251129/population-in-china-by-gender/].

The Australian Seniors Series (2018) 'The modern dating report.' Australian Seniors and CoreData Dating Survey, 15th instalment of national survey on attitudes and concerns of Australians over 50, November [https://www.seniors.com.au/news-insights/discover/australian-seniors-series-modern-dating].

The Australian Seniors Series (2016) 'Seniors and relationships.' Australian Seniors and CoreData Dating Survey, 4th instalment of national survey on attitudes and concerns of Australians over 50, October [https://www.seniors.com.au/news-insights/seniors-and-relationships].

The Equality Trust (2019) 'The scale of economic inequality in the UK' [https://www.equalitytrust.org.uk/scale-economic-inequality-uk#:~:text=UK%20Income%20Inequality,compared%20to%20other%20developed%20countries.&text=In%202018%2C%20households%20in%20the,top%2020%25%20had%20%C2%A369%2C126].

The Telegraph—Dating (2020) 'Older dating: Find love later in life', 24 November [https://www.telegraph.co.uk/online-dating/mature-dating/senior-dating/].

Thompson, L. (2018) '"I can be your Tinder nightmare": Harassment and misogyny in the online sexual marketplace.' *Feminism and Psychology* 28(1): 69–89.

Thompson, L. (2016) '#DickPics are no joke: Cyber-flashing, misogyny and online dating.' *The Conversation*, February 3 [https://theconversation.com/dickpics-are-no-joke-cyber-flashing-misogyny-and-online-dating-53843].

Thompson, T. (2014) 'Queer lives in Fiji.' Unpublished PhD thesis, Sociology, University of Auckland [https://apo.org.au/sites/default/files/resource-files/2013-12/apo-nid110826.pdf].

Thomson, A. Carville, O. and Lanxon, N. (2020) 'Match opts to keep race filter for dating as other sites drop it.' *Bloomberg/Quint*, June 8 [https://www.bloombergquint.com/technology/dating-apps-debate-race-filters-as-empowering-or-discriminating].

Tiffany, K. (2020) 'Tinder's most notorious men.' *The Atlantic*, February 12 [https://www.theatlantic.com/technology/archive/2020/02/tinder-urban-legends-repeat-profiles-kitchen-guy/606406/].

Timerman, J. (2018) 'Misogyny, femicide and an unexpected abortion.' *The New York Times*, May 15 [https://www.nytimes.com/2018/05/15/opinion/argentina-abortion-ni-una-menos.html].

Toglia, M. (2016) 'Figure out which dating site you should try in '16.' *Bustle*, January 1 [https://www.bustle.com/articles/132879-which-dating-app-or-site-should-you-try-in-2016].

Toma, C. L., Hancock, J. T. and Ellison, N. B. (2008) 'Separating fact from fiction: An examination of deceptive self-presentation in online dating profiles.' *Personality and Social Psychology Bulletin* 34(8): 1023–1036.

Townsend, L. and Wallace, C. (2016) 'Social media research: A guide to ethics' [https://www.gla.ac.uk/media_487729_en.pdf].

Træen, B., Carvalheira, A., Kvalem, I. L. and Hald, G. M. (2018) 'European older adults' use of the Internet and social networks for love and sex.' *Cyberpsychology: Journal of Psychosocial Research on Cyberspace 12*(3): Article 1. [https://doi.org/10.5817/CP2018-3-1].

Transparency International (2017) 'Uganda' [https://www.transparency.org/en/countries/Uganda].

Trivers, R. L. (1972/2017) 'Parental investment and sexual selection', pp. 136–179. In B. Campbell (ed), *Sexual Selection and the Descent of Man, 1871–1971*. Abingdon, Oxon and New York: Routledge.

Turkle, S. (1995) *Life on the Screen: Identity in the Age of the Internet*. London: Wiedenfield and Nicolson.

Tyson, G., Perta, V. C., Haddadi, H. and Seto, M. (2016) 'A first look at user activity on Tinder.' IEEE/ACM International Conference on Advances in Social Networks Analysis and Mining (ASONAM), San Francisco, CA. [https://www.eecs.qmul.ac.uk/~tysong/files/Tinder.pdf].

Uganda Communications Commission (2019) blog, 'Communication sector report—September 2019' [https://uccinfo.blog/2020/01/30/communication-sector-report-september-2019/].

Uhlenberg, P. (2009a) 'Introduction', pp. 1–4. In P. Uhlenberg (ed.), *International Handbook of Population Aging*. Berlin/Heidelberg: Springer Science + Business Media.

Uhlenberg, P. (ed.) (2009b) *International Handbook of Population Aging*. Berlin/Heidelberg: Springer Science + Business Media.

UK Parliament (2015) 'Challenges of an ageing population.' Key Issues for Parliament 2015 [https://www.parliament.uk/business/publications/research/key-issues-parliament-2015/social-change/ageing-population/].

UN (United Nations) Development Program GII (2019) 'Human Development Reports: Gender Inequality Index (GII)' [http://hdr.undp.org/en/content/gender-inequality-index-gii].

Urban Dictionary (no date) Definition of 'queerplatonic' [https://www.urbandictionary.com/define.php?term=queerplatonic].

Usher, N., Holcomb, J. and Littman, J. (2018) 'Twitter makes it worse: Political journalists, gendered echo chambers, and amplification of gender bias.' *The International Journal of Press/Politics* 23(3): 324–344.

Uysal, S. (2019) 'Jury out on whether Turks like offline vs online dating.' *Anadolu Agency (AA)*, 17 August [https://www.aa.com.tr/en/life/jury-out-on-whether-turks-like-offline-vs-online-dating/1558330].

Valentine, G. (2006) 'Globalizing intimacy: The role of information and communication technologies in maintaining and creating relationships.' *Women's Studies Quarterly* 34(1/2): 365–393.

Valmorbida, E. (2018) *The Madonna of the Mountains*. London: Faber & Faber.

Vandeweerd, C., Myers, J., Coulter, M., Yalcin, A. and Corvin, J. (2016) 'Positives and negatives of online dating according to women 50+.' *Journal of Women & Aging* 28(3): 259–270.

Vauclair, C-M., Hanke, K., Huang, L-L. and Abrams, D. (2017) 'Are Asian cultures really less ageist than Western ones? It depends on the questions asked.' *International Journal of Psychology* 52(2): 136–144.

Verdenpressen (2017) 'Danish values.' Copenhagen Language Center, 4 August [https://www.kbh-sprogcenter.dk/en/blog/danish-values/].

VicHealth (2003) 'Australian tobacco control legislation.' VicHealth Centre for Tobacco Control, Cancer Council Victoria, for The Australian Council on Smoking and Health (ACOSH), October [https://www.acosh.org/law-policy/australian-tobacco-control-legislation/].

Victor, D (2016) 'Women who make first move in online dating are rewarded, study finds.' *The New York Times*, March 9 [https://www.nytimes.com/2016/03/10/style/women-who-make-first-move-in-online-dating-are-rewarded-study-finds.html].

Victoria Milan Review (2020) 'Victoria Milan Review 2020—Perfect or Scam?' *HookUpWebsites* [https://hookupwebsites.org/victoria-milan-review/].

Victoria Milan (no date) 'About Us' [https://www.victoriamilan.com/en_AU/static/about-us].

Victoria Milan (no date) blog, 'Affair dating cheating saving marriages' [https://www.victoriamilan.com/blog/affair-dating-cheating-saving-marriages/].

Victorian Department of Health (2017) 'Strong relationships, strong health' [https://www.betterhealth.vic.gov.au/health/HealthyLiving/Strong-relationships-strong-health].

Vine, L. (2018) 'I'm addicted to dating apps—but I don't want a date.' *BBC*, 6 April [https://www.bbc.co.uk/bbcthree/article/a0abe5ca-ad4b-4c08-8a10-a616480711ff].

VirtualWayfarer (2015) 'Denmark 101—Dating and Danish Women—Episode 17.' Video, *Facebook*, November 16 [https://www.facebook.com/VirtualWayfarer/videos/10153952069063974].

Visser, R. and McDonald, D. (2007) 'Swings and roundabouts: Management of jealousy in heterosexual "swinging" couples.' *British Journal of Social Psychology* 46(2): 459–476.

Visual.ly (2020) 'UK online dating stats', May 31 [https://visual.ly/community/Infographics/love-and-sex/uk-online-dating-stats-dating-friends?action=likes].

Viveros, J. and Schramm, D. (2018) 'Pros and cons of online dating in later life', April. Families and Communities, Extension, Utah State University [https://digitalcommons.usu.edu/extension_curall/1865/].

Vogels, E. (2020) 'Ten facts about Americans and online dating.' Pew Research Center, February 6 [https://www.pewresearch.org/fact-tank/2020/02/06/10-facts-about-americans-and-online-dating/].

Von Scheve, C. (2018) 'A social relational account of affect'. *European Journal of Social Theory* 21(1): 39–59 [https://doi.org/10.1177/1368431017690007].

Wada, M., Mortenson, W. B. and Clarke, L. H. (2016) 'Older adults' online dating profiles and successful aging.' *Canadian Journal on Aging* 35(4), December: 479–490.

Wahling, A., Duncan, D., Angelides, S. and Dowsett, G. (2018) 'Men and masculinity in men's magazines: A review.' *Sociology Compass* 12(7): 1–26.
Wajcman, J. (2007) 'From women and technology to gendered technoscience.' *Information, Community and Society*' 10(3): 287–298.
Wajcman, J. (2004) *TechnoFeminism*. Cambridge: Polity Press.
Waldi, B. and Byron, R. (eds) (2006) *Migration and Marriage: Heterogamy and Homogamy in a Changing World*. Berlin and Vienna: Lit Verlag.
Waling, A. and Pym, T. (2019) ' "C'mon, no one wants a dick pic": Exploring the cultural framings of the "Dick Pic" in contemporary online publics.' *Journal of Gender Studies* 28(1): 70–85.
Walker, L., Butland, D. and Connell, R. W. (2000) 'Boys on the road: Masculinities, car culture, and road safety education.' *The Journal of Men's Studies* 8(2), March 1: 153–169.
Walker, L. J. and Heard, G. M. (2015) 'Interethnic partnering: Patterns by birthplace, ancestry and indigenous status', pp. 53–75. In G. Heard and D. Arunachalam (eds), *Family Formation in 21st Century Australia*. Dordrecht, Netherlands: Springer.
Walker, L. J. and Heard, G. (2012) 'Population change: Past, present and future.' *The 16th Biennial Conference of the Australian Population Association*, Melbourne, Victoria, 5–7 December.
Wallace, A., Kurti, L. and Tomiczek, C. (2017) 'The drivers of ageism.' Report by The Benevolent Society, September.
Walton-Roberts, M. (2010) 'The family and field work: Intimate geographies and countertopographies, pp. 170–183. In B. Hallman (ed.), *The Spatiality of Families and Family Life*. Oxford: Oxford University Press.
Wamala, C. V. (2013) 'I have to give an "I can" attitude.' *SAGE Open*, *3*(1); DOI: 10.1177/2158244013477101.
Wang, X. and Nehring, D. (2013) 'Individualization as an ambition: Mapping the dating landscape in Beijing.' *Modern China* 40(6): 578–604.
Warner, M. (1999) *The Trouble with Normal: Sex, Politics and the Ethics of a Queer Life*. Cambridge, MA; Harvard University Press.
Watts, T. (2019) *The Golden Country: Australia's Changing Identity*. Melbourne: Text Publishing.
Weber, R. P. (1990) *Basic Content Analysis*, 2nd ed. Newbury Park, CA: Sage.
Weeks, D. (2002) 'Sex for the mature adult: Health, self-esteem and countering ageist stereotypes.' *Sexual and Relationship Therapy* 17(3): 231–240.
WEF (2020) 'Global Gender Gap Report 2020—Global Results.' Geneva, Switzerland: World Economic Forum [http://www3.weforum.org/docs/WEF_GGGR_2020.pdf].
WEF Performance by Region (2020) 'Global Gender Gap Report 2020.' [https://reports.weforum.org/global-gender-gap-report-2020/the-global-gender-gap-index-2020/performance-by-region-and-country/].
Weitz, R. (2010) 'Changing the scripts: Midlife women in contemporary U.S. film.' *Sexuality & Culture* 14: 17–32.
Welzel, C., Ingelhart, R. and Kilgemann, H-D. (2003) The theory of human development: A cross-cultural analysis.' *European Journal of Political Research* 42(3), May; 341–379.
West, C. and Zimmerman, D. (1987) 'Doing gender.' *Gender & Society* 1(2): 125–151.
Whitbourne (2017) 'The secret reason why sex is so crucial in relationships.' *Psychology Today*, July 4 [https://www.psychologytoday.com/us/blog/fulfillment-any-age/201707/the-secret-reason-why-sex-is-so-crucial-in-relationships].

White, H., McConnell, E., Clipp, E., Bynum, L., Teague, C., Navas, L., Craven, S. and Halbrecht, H. (1999) 'Surfing the Net in later life: A review of the literature and pilot study of computer use and quality of life.' *Journal of Applied Gerontology* 18(3): 358–378.

White, M. (2005) 'Children, trauma and subordinate storyline development.' *International Journal of Narrative Therapy and Community Work* 3&4: 10–23.

White, N. (2019) '"My soulmate is Black": Why race really matters to 20-somethings when dating online.' *Huffington Post*, 16 May [https://www.huffingtonpost.co.uk/entry/race-online-dating_uk_5cdc4030e4b066205c601d51].

Whitty, M. T. (2008a) 'Liberating or debilitating? An examination of romantic relationships, sexual relationships and friendships on the net.' *Computers in Human Behavior* 24(5): 1837–1850.

Whitty, M. T. (2008b) 'Revealing the "real" me, searching for the "actual" you: Presentations of self on an internet dating site.' *Computers in Human Behavior* 24(4): 1707–1723.

Whitty, M. T. (2008c) 'The art of selling one's "self" on an online dating site: The BAR approach', pp. 57–69. In M. T. Whitty, A. J. Baker, and J. A. Inman (eds), *Online Matchmaking*. Basingstoke: Palgrave Macmillan.

Whitty, M. T., Baker, A.J. and Inman, J.A. (eds) (2007) *Online Matchmaking*. Basingstoke: Palgrave Macmillan.

Whyte, S. and Torgler, B. (2017) '"Things change with age": Educational assortment in online dating.' *Personality and Individual Differences* 109: 5–11.

Wilding, F. (1998) 'Where's the feminism in cyberfeminism?' *n. paradoxa: international feminist art journal* 1(2): 6–13.

Wilkinson, E. (2012) 'The romantic imaginary: Compulsory coupledom and single existence.' In S. Hines and Y. Taylor (eds), *Sexualities: Past Reflections, Future Directions. Genders and Sexualities in the Social Sciences*. London: Palgrave Macmillan [https://doi.org/10.1057/9781137002785_8].

Williams, G. (1999) 'The federal parliament and the protection of human rights.' Research Paper 20 1998–99, Parliament of Australia [https://www.aph.gov.au/About_Parliament/Parliamentary_Departments/Parliamentary_Library/pubs/rp/rp9899/99rp20].

Willis, M., Jozkowski, K. N., Lo, W. and Sanders, S. A. (2018) 'Are women's orgasms hindered by phallocentric imperatives?' *Archives of Sexual Behaviour* 47: 1565–1576.

Winsor, B. (2014) 'Denmark is bigger than you ever imagined.' *Business Insider Australia*, October 4 [https://www.businessinsider.com.au/denmark-is-unbelievably-massive-2014-10].

Wion, R. K. and Loeb, S. J. (2015) 'Older adults engaging in online dating: What gerontological nurses should know.' *Journal of Gerontological Nursing* 41(10): 25–35.

Wong, W. M. (2014) 'Finding "love" in China: An overview of Chinese marriage markets.' *Inquiries Journal/Student Pulse* 6(12), Social Sciences, Arts, & Humanities: 1–3.

Workplace Gender Equality Agency (2020) 'Australia's gender pay gap statistics 2020.' Canberra, Australia.

World Bank (2017) 'Individuals using the internet (% of population)—Uganda.' [https://data.worldbank.org/indicator/IT.NET.USER.ZS?locations=UG].

World Bank (2016) 'The Uganda poverty assessment report: Farms, cities and good fortune—assessing poverty reduction in Uganda from 2006 to 2013.' Report No.

ACS18391, September [http://pubdocs.worldbank.org/en/381951474255092375/pdf/Uganda-Poverty-Assessment-Report-2016.pdf].
World Health Organization (WHO) (2006) 'Defining sexual health: Report of a technical consultation on sexual health', 28–31 January 2002, Geneva [http://www.who.int/reproductivehealth/publications/sexual_health/defining_sexual_health.pdf].
World Health Organization (WHO) (2002) *Active Ageing: A Policy Framework*. WHO, Europe.
World Health Organization (WHO) (no date) 'Ageing and life-course: Frequently asked questions' [https://www.who.int/ageing/features/faq-ageism/en/].
World Values Survey Wave 7 2017–2020 (2020) 'WVS Wave 7 (2017–2020)' [http://www.worldvaluessurvey.org/WVSDocumentationWV7.jsp].
Worldometers (2020) 'Uganda demographics 2020' [https://www.worldometers.info/demographics/uganda-demographics/].
Worldometers (2020) 'Uganda population' [https://www.worldometers.info/world-population/uganda-population/#:~:text=The%20current%20population%20of%20Uganda,year%20according%20to%20UN%20data].
Worldometers Australia Demographics (2020) 'Australia demographics' [https://www.worldometers.info/demographics/australia-demographics/].
Worldometers Life Expectancy by Country (2020) 'Countries ranked by life expectancy' [https://www.worldometers.info/demographics/life-expectancy/].
Worldometers worldwide life expectancy (2020) 'Life expectancy of the world population: Countries ranked by life expectancy' [https://www.worldometers.info/demographics/life-expectancy/#countries-ranked-by-life-expectancy].
Worldpopulationreview Countries by Area (2020) 'Largest countries in the world 2020' [https://worldpopulationreview.com/country-rankings/largest-countries-in-the-world].
Worldpopulationreview Population by Country (2020) '2020 world population by country' [https://worldpopulationreview.com/].
Worldpopulationreview (2020) 'Total population by country' (UK, Uganda, Denmark, Australia, Argentina, Turkey, China) [https://worldpopulationreview.com/countries].
Worldpopulationreview UK-cities (2020) 'Population of cities in United Kingdom, 2020' [https://worldpopulationreview.com/countries/cities/united-kingdom.
Wouters, C. (2008) *Informalization: Manners and Emotions Since 1890*. London: Sage.
Wouters, C. (2004) 'Changing regimes of manners and emotions: From disciplining to informalizing', pp. 193–211. In S. Loyal and S. Quilley (eds), *The Sociology of Norbert Elias*. Cambridge: Cambridge University Press.
Wouters, C. (1995a) 'Etiquette books and emotion management in the 20th century: Part One: The integration of social classes.' *Journal of Social History* 29(1): 107–124.
Wouters, C. (1995b) 'Etiquette books and emotion management in the 20th century: Part Two: The integration of the sexes.' *Journal of Social History* 29(2): 325–39.
Wu, H. (2018) 'Homogamy in U.S. marriages, 2016.' *Family Profiles, FP-18–18*. Bowling Green, OH: National Center for Family & Marriage Research [https://www.bgsu.edu/ncfmr/resources/data/family-profiles/wu-homogamy-marriages-fp-18-18.html].
Wyrod, R. (2009) 'Between women's rights and men's authority: Masculinity and shifting discourses of gender difference in urban Uganda.' *Gender & Society* 22(6): 799–823.
Yan, Y. (2010) 'The Chinese path to individualization.' *British Journal of Sociology* 61(3): 489–512.

Yancey, G. (2009) 'Cross racial differences in the racial preference of potential dating partners.' *The Sociological Quarterly* 50(1): 121–143.

Yancey, G. (2007) 'Homogamy over the net: Using internet advertisements to discover who interracially dates.' *Journal of Social and Personal Relationships* 24(6): 913–930.

Ybarra, M, L., Kiwanuka, J., Emenyunu, N. and Bangsberg, D. R. (2006) 'Internet use among Ugandan adolescents: Implications for HIV intervention.' *PLOS Medicine* 3(11), November 7: e433 [https://www.ncbi.nlm.nih.gov/pmc/articles/PMC1630714/].

YouGov (2017a) '43% of Chinese Nationals have used internet dating.' [https://china.yougov.com/en-cn/news/2017/11/23/internet-dating/].

YouGov (2017b) 'A third of Aussies have used internet dating.' [https://au.yougov.com/news/2017/11/23/internet-dating/].

Yurchisin, J., Watchravesringkan, K. and McCabe, D. B. (2005) 'An exploration of identity recreation in the context of ID.' *Social Behavior and Personality: An international Journal* 33(8): 735–750.

Zelaya, I. (2020) 'Dating apps speak out against racism and reckon with ethnicity filters.' *AdWeek*, June 2 [https://www.adweek.com/brand-marketing/dating-apps-racism-ethnicity-filters/].

Zytko, D., Grandhi, S. and Jones, Q. (2018) 'The (un) enjoyable user experience of online dating systems', pp. 61–75. In M. Blythe and A. Monk (eds), *Funology 2*. Cham, Switzerland: Springer.

Index

For Product Safety Concerns and Information please contact our EU representative GPSR@taylorandfrancis.com
Taylor & Francis Verlag GmbH, Kaufingerstraße 24, 80331 München, Germany

www.ingramcontent.com/pod-product-compliance
Ingram Content Group UK Ltd.
Pitfield, Milton Keynes, MK11 3LW, UK
UKHW022352190926
475702UK00015B/2252

* 9 7 8 0 3 6 7 7 5 3 9 3 1 *